"1969: Moon Landing? A Conspiracy Theory Investigation"

by IBM

First edition, 2023 by IBM ibm
publishingh

This is a book about the doubts and inconsistencies surrounding the historical event of the Apollo moon landing, which the author argues may have been a "virtual" reality created through technical and cinematographic means, as well as social engineering. The author presents a variety of logical, political, and social clues, as well as factual evidence, to support this argument. From the alleged flag waving in a vacuum to the lack of stars in the photographs, this book analyzes and questions the commonly accepted narrative of the moon landing.

Through careful examination of official documents, technical specifications, and eyewitness accounts, the author presents a compelling argument for the possibility of a moon landing conspiracy. This book is a must-read for anyone interested in the history of space exploration and the mysteries surrounding the 'so called' moon landing.

Summary

1. Introduction: the Truth

I don't care who believes it or not. I probably won't benefit from this work financially, and I'm sure that none of the many conspiracy theorists who have written books on this topic have become particularly wealthy or famous. What really interests me is to ensure that people don't take at face value what is offered to them by the mainstream.

If Columbus had listened to the scholars of his time, he would never have set sail. "And yet he moves," Galileo exclaimed in front of the Inquisition, recanting his celestial observations. Of course, we shouldn't question incontrovertible data, such as the efficacy of vaccines (proven by decades of practice) or the fact that the earth is round, which can be easily observed from a scheduled flight at a sufficient altitude. However, it's important to instill doubt about whether we really know what we've been told and whether it can be verified except through faith in NASA, astronauts, and scientists who have told us.

Because everything we saw on TV, from 1969 to 1972, can be effectively falsified. Not only that, it was extremely unlikely that Americans or anyone else could have really landed on the moon. 50 years after the landing on 20 July 1969, doubts surrounding the moon landing are fueled by a series of photographs and official NASA videos full of technical inconsistencies.

This book analyzes the obscure points connected to the recent history of the conquest of our satellite, examining the anomalies found in the photographs of the time and evaluating possible alternative scenarios, from a reconstruction in a film set to the possible retouching of some photographs to hide something that should not be shown to the public.

This book aims to present all the logical and/or scientific arguments to show that the historical event known as "the landing of man on the moon" can be a "virtual" reality, built at a table and realized with technical/cinematographic means, as well as with a masterful work of social engineering that has manipulated people's minds to the point of making the United States' "conquest of the Moon" a dogma.

Of course, there is no "legal" evidence of what has been affirmed, but there are a myriad of logical, political, and social clues, as well as tens of factual pieces of evidence relating to physical, sound, and optical phenomena that are inexplicable in any other way than by tampering with or creating ex novo "evidence" of the Apollo landing.

There are also important omissive clues in the event of 20 July 1969, such as the resignation of NASA's top management a few months after the epochal outcome and, above all, the loss of the original registration of the integral data of the lunar mission.

This book takes advantage of all these facts as narrated in various literary and filmic works related to the so-called "lunar conspiracy," most of which come from research carried out for the most recent of them, "American Moon," by Massimo Mazzucco.

The book is divided into 17 chapters (like the Apollo!) that address the mission of Apollo 11 from its conception in 1961, by an extemporaneous declaration by the then-President J.F. Kennedy. The story then unfolds between the policy that wishes at all costs to affirm the US supremacy and technology, represented by NASA and its Director, James Webb, who sees presidential ambitions as an excellent springboard for personal career and to make the entity directed by him the most important agency of the country after the CIA (from which it derives).

We then move on to the real technical disasters that occurred in the attempt to reach the goal set, even after the assassination of Kennedy and with the confirmations of the project by Johnson and Nixon. The enormous sums pumped by the State to NASA, the myriad of contracts and subcontracts scattered throughout the country make the Apollo project an important economic driver for the USA.

In 1967, however, two years before the so-called moon landing, NASA leaders unexpectedly resigned, including the director and the astronaut in charge of being the first man on the moon. Since then, the inconsistencies in the Apollo project have multiplied, as have the technical failures and real tragedies. Moreover, the form to physically arrive on the Moon seems to be, in reality, a shabby and precarious hut that needs a myriad of redesigns. Immediately after the so-called moon landing, which was broadcast worldwide and seen by over two billion people, the first doubts arise and multiply. Did director Stanley Kubrick play a role? And what about the powerful Walt Disney movie structure? The former had given birth to the movie '2001: A Space Odyssey' a few months before the landing, while the latter had created 'Future World West' in 1966, a world of the future enclosed in a gigantic area of Florida, a direct descendant of the four pavilions created for the 1962 Seattle Expo with the explicit title: 'The Man in the Space Age'. The technical know-how for the visual representation of the lunar conquest therefore existed, and since the beginning of the Cold War, several American scholars had been developing mass manipulation techniques that will have very solid examples in the experiment on the Conformism of Solomon Asch or in manipulations of the American people already implemented in the 1950s by Edward Louis Bernays on behalf of the CIA. However, the criticisms of the official thesis of the "lunar conquest" did not fail to appear. In 1974, two years after the conclusion of the last Apollo mission, Bill Kaysing, a writer from Chicago, published at his own expense the book 'We Never Went to the Moon - America's 30 Billion Dollar Swindle', in which he presented a long series of alleged trials of counterfeiting by US space companies.

Following that initial intuition, today there are millions of individuals all over the world who have decided not to accept NASA's thesis and, with the scientific evidence available, have rejected the idea of the 1969 landing as a dogma of faith. This book is dedicated to all those who are not satisfied with simple answers.

Apollo, son of Zeus and Leto, was the god of music, poetry, prophecy, and healing in Greek mythology. He was also associated with the sun and archery, and was considered the leader of the Muses. In Roman mythology, he was often identified with the god of the same name, but was also associated with the god of war, Mars. The myth of the moon, on the other hand, has been present in various cultures and mythologies throughout history. The propaganda of past victories increases the strength of the spirit and the pride of the nation, unites it and creates the basis for victories on all fronts of geopolitical battles, in the present and in the future.

Alexander Ivanovich Popov, author of the book "Americans on the Moon: A Great Breakthrough or Space Scam?" (Moscow, 2009)

If the astronauts, once they reached the Moon in good health, had had no way to return to Earth and died of hunger or thirst, it would have been a catastrophe.

Ernst Stuhlinger, collaborator of Wernher Von Braun at NASA

Brainwashing is the technique of reducing the subject's ability to think independently or critically, in order to introduce new, unwanted thoughts into their mind, as well as to change their habits, values, and beliefs. (Source: Wikipedia)

The act of searching for the truth has always had varying fortunes throughout history. Socrates was put to death by the authorities, accused of subverting the morals of the youth. Galileo was forced to abjure a truth that was evidently true, in order to avoid being imprisoned in some papal fortress and silenced. Assange was arrested for disclosing secrets that interested the community and should have been disclosed by the same authorities who deprived him of his freedom. Many significant events in human history have been hidden by someone who would benefit from the secret or would be damaged by its publicity. One of these events, the most important for its technological implications and the affirmation of our species and a particular country, is the conquest of the Moon. This book does not aim to prove that humans have never been to the Moon, nor does it intend to do so. However, what it can demonstrate is that it is not possible to accept the story told to us as unquestionable simply because there are dozens of physical proofs and an innumerable series of indications that suggest that that event could have been created on Earth by someone interested in writing history for their own use and consumption.

That someone, in this case, would be the United States of America, or better, its successive governments from 1901 until today. The history of the Moon and its fictional story began in 1961, at the height of the Cold War between the US and the USSR, due to the enormous ambitions and unacknowledged fears of an American president who was more mythologized abroad than at home: John Fitzgerald Kennedy.

2. Get to the Moon

When, at 7:28 pm on October 4th, 1957, Nikita Khrushchev's Soviet Union launched Sputnik 1, the first artificial satellite to orbit the Earth, many in Washington were sweating. The first, of course, was the US President at the time, Dwight David Eisenhower, who, as a career soldier, immediately understood the war implications of the Soviet success. The USSR had shown that it possessed the necessary means and technologies to bring nuclear warheads into orbit and strike the United States without warning.

Just one month later, on November 3, 1957, the USSR sent the first living being, the little dog Laika, into orbit. The USA could not respond to Soviet technological superiority until the beginning of 1958, when they succeeded in launching Explorer 1. The satellite proved to be quite important for the purposes of the lunar conspiracy thesis, since it discovered the Van Allen Belts, which would be the stone guest of American space expeditions.

To meet the "red" threat, Eisenhower decided to create an ad hoc government agency to fight the space war. On July 29, 1958, NASA, the National Aeronautics and Space Administration, was born, replacing NACA, the National Advisory Committee for Aeronautics, founded at the dawn of the aviation sector in 1915.

The fight and response in the space race alternated at that point between the two superpowers, with the USSR managing to get a probe to the Moon on September 13, 1959, until the fateful date of April

12, 1961, when the Soviets flew the first man in space, Yuri Gagarin, making him orbit for 00 minutos around the planet.

At this point, the United States was almost in a panic. Although they had captured the brain of the Nazi V2 missile father, Wernher Von Braun, they were losing the space race and were the eternal second to the greatest communist power in the world. It was an unacceptable situation.

Eisenhower was no longer the President of the United States and was succeeded by John Fitzgerald Kennedy, the enterprising and ambitious son of a controversial Boston billionaire who admired Adolf Hitler. Just eight days after the Soviet triumph, Kennedy instructed his deputy, Johnson, to find out if there were any other space programs with phenomenal results in which the US could win. Johnson naturally turned to the only person who could answer this question: the director of NASA, James E. Webb. The answer to Kennedy came after just another eight days: among the various possibilities, there was the exploration of the Moon by a human being that "covers not only an enormous propagandistic interest but is an essential goal whether we are the first to obtain this result or not. But we can be the first. We cannot miss such a goal as it will be a source of knowledge and experience for even greater successes in space." Only a few weeks later, on May 25, 1961, Kennedy launched the space gauntlet to the Soviets, asking Congress for the funds needed to "bring a man to the moon and bring him back safe and sound by the end of the decade. The time has come to accelerate to complete a great new American company! This nation must have a leading role in conquering space."

It was from this discourse that the Apollo program was born, or rather, the reconversion of it started as a continuation of the Mercury and Gemini orbital programs and a real project of conquest of the Moon. The first budget of the new space program was unanimously voted by the Senate. Funds for NASA went from $500 million in 1960 to $5.2 billion in 1965. In total, the funds foreseen to bring the man to the moon by 1970 were the staggering figure of 22 billion dollars. The final balance would reach over $135 billion. In practice, NASA had the budget of a small nation available in ten years.

3. The cold space war

According to Jacques Villain's book "À la conquête de l'espace: de Spoutnik à l'homme sur Mars" published in 2007, at the beginning of his presidency, Kennedy did not intend to invest huge resources in space exploration. However, due to repeated Soviet successes and the need to regain credibility after the failed Bay of Pigs invasion, aimed at ousting Fidel Castro, space exploration became indispensable.

The US managed to send Alan Shepard into space a month after Kennedy's speech, but only for a suborbital flight. To match the USSR's achievement of John Glenn's orbital flight, they had to wait for almost another year, until February 20, 1962. Meanwhile, the Soviets continued to progress rapidly. In August 1962, they made the first flight with two spaceships at the same time. On June 16, 1963, they sent Valentina Vladimirovna Tereškova into Earth's orbit, becoming the first woman to achieve this feat.

The first shuttle with three cosmonauts on board, Voschod 1, was launched on October 12, 1964. This flight was also the first in which the crew did not wear spacesuits. Aleksej Archipovič Leonov, of Voschod 2, launched on March 18, 1965, also carried out the first spacewalk in history, although there was a risk of tragedy due to a problem with the retrorockets that landed the shuttle about 1600 km from the predetermined site.

Meanwhile, the Apollo project continued to progress. NASA's director, Webb, was making every effort to please his president and deserve the huge funds allocated to him. He focused all NASA's resources on the Moon mission, reorganizing it for this unique purpose.

In the face of growing Soviet successes, Kennedy summoned Webb to the White House on September 18, 1963, just over two months before his assassination in Dallas, to be reassured about the feasibility of the impossible project. "If I am re-elected, we will not have time to go to the Moon at this time, will we?" the President asked worriedly. "No, we won't make it, but we'll probably fly around it," replied Webb contritely. Here, Kennedy seemed to falter as he asked a question that should have been asked before his firing in Congress: "Do you think it is a good idea to send a man to the Moon?" "Certainly," replied Webb, probably thinking of the enormous funds given to his institution. The President, however, urged him: "And wouldn't it cost much less to go there with probes?" "No, it's not the same thing," said Webb. "The lunar program has given us the impetus to build huge carrier rockets for this specific purpose and will only work if built for this purpose." Kennedy seemed discouraged and perhaps even frightened: "To me, it seems like a mountain of money to go to the Moon when we can find out scientifically, with instruments, almost everything that interests us. Sending a man to the Moon is just a circus number that is not worth all those billions of dollars."

On September 20, two days after the dramatic confrontation with Webb, Kennedy made a speech at the UN, attempting to involve the Soviets in the conquest of the Moon by asking for technical-scientific collaboration in space exploration. He proposed a joint effort for exploration and a two-way regulation of the new frontier that cuts off the rest of the world. Kennedy even explicitly asked the Soviets for a joint expedition to the Moon.

Kennedy's attempt to save himself from a gigantic financial hemorrhage in his own country was shipwrecked miserably with Nikita Khrushchev's reply, which arrived the following month, on October 26th: "At the present time, we have no plans to send cosmonauts to the moon. I read that the Americans intend to do it by 1970. We wish them to succeed. And we'll see how they fly up there and how they land or, better said, how they will align. And above all, we will see how they will start and come back. We do not intend to compete in sending people to the moon without adequate preparation."

Khrushchev's reply was a poisoned meatball. At this point, Kennedy should have carried out the Luna project alone. If only he had been alive to do it. Less than a month later, on November 22nd, his head was shot off on Elm Street in Dallas aboard a huge Ford Lincoln convertible built especially for him. A few hours before he died, he proclaimed: "This country has thrown its heart beyond the obstacle of space, and we have no choice but to follow it."

To the new President of the United States, Lyndon Johnson, there was nothing left but to keep faith with the irrevocable commitment that Kennedy had taken before the whole world.

Meanwhile, the Soviets were actually studying the possibility of getting a man to the Moon but, even in their case, fate got in the way. In January 1966, the great architect of the Soviet space adventure, the engineer, died suddenly. Sergei Pavlovic Korolev (known only as 'the Chief Designer' to protect him from possible attacks by American secret services), was the one who had literally invented the rockets of the USSR. His was the vector R-7 "Semërka," the first intercontinental ballistic missile in history, the one that put Sputnik and the little dog Laika into orbit.

The father of Soviet space remained under the knife because of an unstoppable hemorrhage during a surgical operation to remove a tumor of the intestine, initially diagnosed as a simple polyp. It can be said that the Soviet lunar adventure ceased at that time, although not the spatial one. Certainly, the USSR had lost the fundamental figure from a technical and charismatic point of view from the media point of view that might have allowed it to face the aggressive US 'lunar' campaign.

4. Announced catastrophe

In the weeks that followed, four N1 carriers exploded at the start. They should have been used for the lunar mission.

On the other hand, if the Soviet Union reluctantly gave up its own lunar program, the United States was grappling with real earthly 'aliens', the Vietcong. The Vietnam war was absorbing ever-greater resources, both human and financial, generating ever greater concerns and protests among the American population.

In this context, the Apollo program was now out of control: its costs had breached every budget before a single rocket had been launched, and someone was already airing the word that so much terror inspired in Kennedy and in all American politicians: "failure."

In the autumn of 1966, General Samuel C. Phillips, director of the Apollo project, presented a report that denounced the state of general confusion, uncontrolled expenditure, and technological setbacks of the lunar program and the development of the Saturn V vector with respect to the targets set. In particular, the Phillips Report pointed the finger at the inadequacy of the Apollo command module and the malfunctioning of the second stage of the rocket. These two components had been contracted to the North American Aviation company in Inglewood, California.

After four and a half years, and just over a year and a half after the planned launch, there are still important technical and unknown problems," wrote the General, who continued relentlessly: "There are problems with the electrical system, the secondary propulsion system, the structural hold, and the increase in weight that must still be solved. The delay or incompleteness of the certification programs makes us seriously fear that we will not have fully qualified vehicles to carry out the lunar program."

According to some documents, it seems that the Director of NASA, Webb, did not even learn about the Phillips Report. Still, this did not prevent him from continuing to beat the drum for more funding. In a conversation with Johnson on October 5, 1966, Webb claimed that the reason for his requests for funds was not frivolous: "These people must send men into space. We cannot go to North American Aviation and tell them that we cut $130 million from their budget, but we expect a lot more work for $130 million less."

In short, the problems highlighted in the Phillips Report continued to multiply. A fuel tank of the control module was seriously damaged during testing, the environmental control system did not work at all, and a complete redesign was necessary. The second stage of the Saturn V exploded twice in 1965 and '66, greatly delaying the schedule of the Apollo program. Moreover, during that period, cracks were discovered in the structure of the rocket and in the fuel tanks.

Even the construction of the LEM, the real landing module, was barely progressing. Every day, something was found that had been badly built or gave problems because it had been poorly installed. There were broken or almost broken electrical cables, fuel leaks in every ring of the Apollo system. With all these technical failures, we reach the fateful year of 1967. A key year for the Apollo affair.

In January, Thomas Ronald Baron, one of the North American Aviation security inspectors, presented a report of about fifty pages confirming all the problems highlighted by General Phillips. In the Baron report, he stressed the multiplicity of accidents that occurred due to the lack of preparation by construction workers, the lack of spare parts, procedures, and very little attention to safety procedures. Lack of coordination between the program's sectors, any communication between anyone, sector managers who did not take problems seriously, technicians who did not know their job, lack of respect for delivery times, and even very little cleaning of the workplace. The Baron report was merciless and concluded by writing that the safety of astronauts should not be jeopardized just to fulfill a planned work program.

The Baron report was like a Cassandra prophecy: it was ignored by NASA, and Baron was fired. A few days after its disclosure, on January 20, the Saturn V carrier exploded during the testing phase. But seven days later, what happened could have written the definitive end to US ambitions on the conquest of the Moon. During a simple simulation on the ground of the Apollo 1 shuttle, the first experience with Saturn V and men on board, the three astronauts Virgil Grissom, Edward White, and Roger Chaffee died on the launch pad, burned alive in atrocious suffering witnessed also by a convulsive and defective radio communication with the control tower. "How are we going to go to the moon if we can't even talk between three buildings?" were Grissom's last words.

Following the tragedy, they began to completely redesign the space shuttle, determining at least a year of delay on the lunar roadmap. Still, in the meantime, due to the Houston massacre, the Apollo program was suspended indefinitely before a single spaceship was launched.

Meanwhile, Baron had released his report to the press, adding more details until it became a 500-page dossier that he delivered directly to the US Congressional committees investigating the disaster. On April 21, 1967, he testified before a subcommittee led by Congressman Olin Teague. One week after testifying, Baron and his family died when their car was struck by a train at a level crossing. His report has not resurfaced since.

4. James Webb and the mission impossible

We always knew that something tragic would happen sooner or later, but who would have thought that the first tragedy would happen on the ground?" The Director of NASA, James Webb, who spoke with the presidents and managed to secure billions from the state budget, reacted to the Apollo 1 drama.

It was a beautiful spring day in the same year when he presented himself to the Parliamentary Science and Astronomy Commission. However, what they found in front of the members of Congress was a very different Webb from the one who nourished Kennedy's delusions of omnipotence. Before the Commission investigating the disaster, he declared, "If we make it by the end of 1969, we will be very, very lucky. The chances of finishing all the necessary work are fewer this year than they were last year, and already a year ago, I had witnessed at this same table that the chances of making it were even less than those of the year before." In short, the closer the expiration date set by Kennedy, the more unlikely it became that the Moon target could be reached.

Just over a year after these words, on October 7, 1968, Webb unexpectedly resigned from NASA. The man who had spent seven years of his life pursuing the dream of the Moon, which had nurtured the hopes of two presidents, who had used all his oratory art to obtain state funds to feed the voracious Apollo project, suddenly retired a few months after reaching the desired goal.

Even more strange was the resignation of Webb's deputy, Robert C. Seamans, a graduate in engineering from Harvard with a master's degree in Aeronautics from MIT in Boston, who retired before Webb, in January 1968, to become an adjunct professor. He had all the credentials to become the new director, instead of Webb. Obviously, his resignation raised questions. The Washington Post called him for an interview in which Seamans declared that he had been at NASA for seven years and, in reality, he wanted to stay only two.

That his resignation had to do with the catastrophic situation of the Apollo project also made other resignations seem suspicious, such as astronaut Walter Schirra's. Schirra, of Sardinian origin, had participated in both the Mercury space program (he had been one of the Mercury 7, the first astronauts of the American space program) and the Gemini program. Schirra was probably the most qualified astronaut for a mission on the Moon, but four days after Webb, on October 11, 1968, shortly before leaving aboard the Apollo 7, he announced that this would be his last mission with NASA.

In orbit, Schirra stayed there for 11 days, and for the first time, live television images were transmitted from inside a spaceship. Surely, Schirra would have been destined to go to the Moon, but instead, on July 1, 1969, 19 days before the Apollo 11 landing, he retired to become a manager in private companies.

Three resignations of this magnitude, all within a few months or even a few days of the "conquest of the Moon", remain to date logically inexplicable

5. The people wanted the Moon

After the Apollo 1 catastrophe, many people within NASA thought that the Luna project should be abandoned once and for all. Chris Kraft, a NASA veteran and director of the Houston space center, said that after the death of the three astronauts who were burned alive, no one at the Space Agency would have continued with the project if the entire nation had not committed itself to pursuing that ambitious achievement. "We would have made fools of ourselves because of the competition with the Soviets."

Another important reason that pushed the government to pursue the lunar landing was the American public opinion, which was not only accepting of the expensive race for space but was also demanding it. Mass culture had deliberately instilled the notion of "space necessity" starting with cartoons like The Jetsons by Hanna and Barbera. The show, which was known in Italy as "I Pronipoti," was created in 1962, a year before Kennedy's announcement, and depicted a futuristic world centered around space. Notably, the head of the family, George, worked in a spaceship factory.

The American government's intense social campaign presented the conquest of the Moon as a matter of course and a cornerstone in the victory against communist ideology. Mass culture was filled with a multitude of Hollywood films and propaganda animations that celebrated the arrival of Americans on the Moon in advance.

Among the most curious titles were "Nude on the Moon" in 1961, which depicted a Moon populated by naked women (perhaps in honor of Kennedy's virtues), "The Mouse on the Moon" in 1963, which portrayed a small state trying to sneak into the race for the Moon between the two superpowers, "Way...Way Out" in 1966, which depicted the US sending a married couple to the Moon to prevent an American astronaut from succumbing to the temptation of having a relationship with the Soviet female crew that managed a lunar base, and "Countdown" in 1967, which featured an astronaut being sent to the Moon without the necessary technology to make him return.

And of course, there is the absolute masterpiece of science fiction, "2001: A Space Odyssey," made by Stanley Kubrick and released in America on April 2, 1968, fifteen months before the alleged moon landing of Apollo 11. After watching Kubrick's film, it was clear to everyone that the technologies to show a likely arrival on the Moon were already there. Moreover, NASA had many images generated by their simulators.

Films were released showing models of the Apollo shuttle or the Lem module that were ready and in perfect working order. The Manned (later Kennedy) Space Center was described as a "reality in the preparation of man's journeys to the Moon." Isaac Asimov, the greatest science fiction writer of the moment, intervened with articles in major periodicals to sanctify "the first step in space," of course, by Americans.

However, the American people were not only convinced of the goodness of the lunar project but also its essential necessity through a real media campaign.

Werner von Braun, a former Nazi scientist who developed the V-1 and V-2 missiles for Germany during World War II, later worked for the US and played a significant role in the development of the Saturn V rocket, which was used in the Apollo missions to the moon. He published a book in 1960 titled "First Man on the Moon," which some have claimed contained a detailed plan for the Apollo 11 mission. Other books, such as "The Exploration of Space" by Willy Ley and "Man in Space" by William E. Burrows, provided further details about the Apollo program. Almost simultaneously with President Kennedy's declaration in 1961, a "Guide to Space for Elementary Teachers" was also published. But perhaps the most significant contribution to the popularization of space travel came from Walt Disney, who built an entire section of Disneyland called Tomorrowland, dedicated to showcasing a vision of the world of tomorrow, including space travel. A show called "Flight to the Moon" was presented there in 1967, and even popular characters like Mickey and Donald had adventures that took them into space.

According to Massimo Mazzucco, author of the film "American Moon," astronauts became the new hero figures in the 1960s, replacing cowboys in American popular culture and symbolizing the country's pursuit of new frontiers. This cultural climate may have created a favorable environment for NASA's moon landing program. However, conspiracy theorists claim that this widespread propaganda made it impossible for NASA to admit that they did not actually land on the moon, despite evidence to the contrary.

6. The lunar 'conspiracy'

William Charles Kaysing was a former United States Navy officer who graduated in English literature and had held various jobs after his discharge. In 1956, he found a job as an editor of technical texts at Rocketdyne, a company based in Canoga Park in the San Fernando Valley that produced rockets for NASA.

A few years later, in 1963, during the height of the Apollo program, he resigned from RocketDyne, possibly due to a nervous breakdown that caused him chronic anxiety and, above all, total disillusionment in technology and the future. During this period, Kaysing got rid of his TV and radio devices and even canceled his magazine subscriptions, leading the whole family into a lifestyle that we would today call nomadic, which lasted almost the entire decade. He lived in a camper van and did all kinds of odd jobs, including picking fruit and ordering tools for dentists.

In 1970, Kaysing gained recognition for two special publications for Paradise Editions: 'The Land and Where to Buy It for a Few Dollars per Acre' and 'Where to Eat Well with Less than a Dollar a Day', which rode the myth of freedom and the new age. The following year, Straight Arrow, the publisher of the already important Rolling Stone magazine since 1967, proposed that he publish his homeless adventures. The resulting book was called 'The Ex-Urbanite's Complete & Illustrated Easy-Does-It First-Time Farmer's Guide', a guide in which he gave practical advice for those who wanted to give themselves to nomadism like him.

In the early 1970s, when the Apollo missions to the moon were already a universally accepted reality, Kaysing began writing something that nobody would have expected. He claimed that during his time as a technical editor at RocketDyne, which had also started producing for the Apollo project, many engineers at the company had argued that perhaps the technology of that time could allow NASA to send a man to the moon but surely not to allow him to restart from there and safely return to Earth. Kaysing revealed that NASA had conducted a feasibility study for lunar missions, which showed there was only a 0.0017 percent chance of success. The scientists quoted by Kaysing also mentioned the powerful radiation of the Van Allen belts, which they considered an important problem to solve before sending humans through them.

In 1976, two years after the last Apollo mission, Kaysing finally gave shape to all these 'revelations' that had evidently been brewing inside him for years. He published at his own expense the book 'We Never Went to the Moon - America's 30 Billion Dollar Swindle', in which he presented a long series of alleged evidence of falsification of US space companies.

"It was simply impossible to go," Kaysing declared in many interviews, "and they were simply inventing a credible alternative for the people."

In his book, Kaysing listed all the reasons why it would not have been possible for a man to have ever arrived on the moon. Some of those observations have now been overcome by plausible explanations, while others certainly have not.

The objections in the book were as follows:

• NASA did not possess the technological capabilities to send a man to the moon and make him turn back.

• The absence of stars in the photographs of the astronauts indicated a scam (as we will see, this 'proof' was then refuted by photography experts: it is not possible for the stars to be seen because of the shutter speed and aperture needed to photograph astronauts and Lem).

• There are inexplicable optical inconsistencies in astronaut photographs.

• There is an inexplicable lack of explosion crater below the lunar module, as the rocket engine of the Lem should have raised a quantity of dust, at least in the last phase of the descent.

• The death of Thomas Baron, the North American Aviation quality inspector, a week after testifying at the Congress about the Apollo 1 tragedy, was too mysterious and a sign of a conspiracy.

According to the book, all the landings had been shot in a secret film studio run by NASA.

Kaysing's epigone was another American, Ralph René, an inventor and amateur physicist. In 1992, he published "Nasa Mooned America", a play on words that in Italian can be rendered with the concept that NASA has made fun of his country. René focused his attention on the Van Allen belts. According to him, the astronauts could not have overcome them without damage, being thousands of kilometers thick and unavoidable to reach the Moon. He also returned to the concept in many interviews that with the means of the 1960s and 1970s, it was impossible for a man to arrive on the moon. "Indeed there is no way to go there today, nor will there be a way tomorrow," he concluded sadly. One of the clues that NASA had never sent men to the moon, according to René, was the fact that they had not conducted experiments with monkeys or other animals before.

Over the years, conspiracy theories have multiplied in parallel with the spread of wider technological knowledge on the part of the people. Besides René and Kaysing, many others have written books on the subject: Bart Bibrel, James Coller, Rick Staddock, Dave McGowan, Gernot Geise, Aron Ranen, David Percy, Takahiko Soejima, Alexander Ivanovich Popov, Mary Bennett, Jarrah White, and Ronnie Stronge, all proposing a reading of the Apollo missions different from the narration of NASA imposed on all humanity.

For an irony of fate, the spread of conspiracy theories has largely been fueled by NASA itself. In fact, since the internet became an essential part of our lives, the American Space Agency has made a significant portion of its Apollo program travel documentation available online. This includes a series of photographs that have drawn the attention of those who believe that it is not possible to have landed on the Moon.

Thanks to this material, more and more people have been able to scrutinize every single frame and discover inconsistencies between what NASA has stated and the photographic and video evidence. Hollywood has also contributed to the debunking of the Moon landing conspiracy. Over the years, many films have portrayed the Apollo mission and its preparation. For instance, in "Countdown," directed by Robert Altman in 1967, NASA sends a man to the Moon without the technology to bring him back home. In 1971, the James Bond film "Diamonds Are Forever" featured a chase scene in a set that resembled the lunar landscape, complete with a Lunar Rover.

The notion that the US government could deceive its citizens was already widespread following the release of the Pentagon Papers in 1971, which showed that the Johnson administration systematically lied about the Vietnam War. Additionally, President Nixon was forced to resign in 1972 due to the Watergate scandal, which further undermined trust in the government's honesty and transparency.

Thus, from the late 1970s, almost a decade after Apollo 11, doubts about the Moon landing began to creep into a series of films. In 1978, "Capricorn One," directed by Peter Hyams, suggested that a mission to Mars should be replaced by a staging to maintain political favor and funding for the space program. The film's poster asked: "Would it upset you to discover that the greatest moment in our history could never have happened?"

The manipulative thesis on the part of NASA also found support in a news episode in that period. In July 1976, the automatic Viking 1 probe landed on Mars and transmitted the first color images of the planet with a brick-red sky. However, NASA published them with a blue sky to avoid upsetting the public, which sparked controversy when discovered.

The certainty of the lunar conspiracy gained new momentum in the new century, a few months before the tragedy of September 11th. On January 18, 2001, the documentary "A Funny Thing Happened on the Way to the Moon" was released by Bart Sibrel. It drew historical parallels with the Tower of Babel and showed the many contradictions in American politics and NASA's accounts of the Moon landing. The documentary argued that America's inferiority in space compared to the USSR could not be tolerated, and the relaunch of the Moon landing was the only card that Kennedy, Johnson, and Nixon could play. The documentary also claimed that the problem of radiation from the Van Allen belts was insurmountable, and that astronauts on the Space Shuttle had already experienced serious safety problems despite being thousands of kilometers from the area.

In February of the same year, Fox then broadcasted a documentary entitled 'Conspiracy Theory: Did We Land on the Moon?', or "Conspiracy Theory: We Landed on the Moon?", which denied all six trips to the Moon between 1969 and 1972, reigniting the theories of Kaysing in 1976. The documentary claimed that if the producer of Capricorn One had made a plausible mission to Mars with a budget of 4 million dollars, then why couldn't NASA have done so with the astronomical figure of $135 billion taken from the pockets of Americans for the entire Apollo program?

But that's not all, because just three years ago in 2015, another film convincingly portrayed the plot of a government conspiracy. It is called "Operation Avalanche" by Matt Johnson. The film is set in 1967 when two CIA agents, who suspect that there is a KGB agent inside NASA with the mission to sabotage the Apollo program, pose as documentary directors and introduce themselves to NASA. Instead of discovering the Soviet spy, they uncover the greatest conspiracy of the century against all humanity.

But why have Hollywood, Disney, and Stanley Kubrick been frequently called into question in the moon landing affair? Simply because the videos of the Apollo missions have many features in common with the best science fiction films of the time, including Kubrick's '2001: A Space Odyssey'. One of the arguments made by supporters of the moon landing is that the sets used for the mission would have been too massive to be plausible, given that NASA has shots even at 360 degrees. However, this objection is easily dismissed by anyone who has studied cinema or seen some films. The technique of inserting real images into immense or imaginary contexts has been used for a long time, even before the advent of digital technologies. Before computer graphics and "green screens," the technique of "front projection" was used. Kubrick himself used it extensively for his science fiction movie (the opening scene of 2001 with the monkeys in the African desert). Only the part of the scene in the foreground is real, and everything behind it is projected onto a rear screen. To determine whether an image exploits this stage trick, one needs to verify the existence of a continuous separation line between the entire scene in the foreground and the rest of the landscape. Unfortunately for proponents of the great lunar enterprise, most of the Apollo mission videos are just like that. A separation line between the background and scene in the foreground is always identifiable, even in 360-degree panoramic views.

The first evidence of the presence of such a technique in the Apollo films was discovered by a writer who was a 'conspiracy' writer from New Mexico, Richard Hoagland, in the 1990s. He took some photos of the Apollo missions and, using the first digital manipulation programs, reduced the contrast and increased the range, discovering the "welds" between the different scenes, the real ones in the foreground and those added in the background.

But the doubts surrounding the moon landing also involve the figure of the man who was the first person to travel to the Moon (with the exception of fantastic literature): Werner Von Braun. The former Nazi scientist had prepared everything for the Apollo spacecraft to put into orbit and then for his journey to the Moon, but in Houston, the organization of the mission and the actual trip was entrusted to Robert Rowe Gilruth, head of the on-board personnel, that is, of the three astronauts. Why did NASA oust Von Braun from the most delicate phase of the mission, that of the landing? In fact, only Gilruth is the true "director" of the landing on the Moon.

His small team, the Space Task Group, consisted of barely 77 people, including technicians and scientists, who controlled all radio communications, live images or images shot or taken by astronauts, and all space operations - including those related to the lunar mission, of course. Gilruth was still responsible for training astronauts who were all military and therefore subject to undisputed obedience and secrecy. The Apollo mission was managed exactly like a military operation because it essentially was one. The Space Task Group, led by Gilruth, was the first nucleus of NASA, which converged in Houston on the occasion of the Apollo missions but originated in Langley, Virginia - a stone's throw from the headquarters of the CIA. Was this just a coincidence?

As Gerard Wisnewski recounts in his book "Lies in Space" from 2005, "Simulations were easy to do. The instructors disconnected the outer stations of downtown Houston (the Manned Space Flight) and linked the data from their simulator data centers. Whoever presided over the mission control center had no way of understanding if the data were true or prevented from simulators. But, besides, he didn't even care."

Do you get it? So the Houston Control Center was not a real control center, but rather a few men who had control of the simulators. One part of the simulator was located in Houston, while the other was in Langley. The media was never informed of these little details of the Apollo missions.

The main text to learn more about space simulation centers (at least eight in the United States) is the book "Spaceflight Revolution," which was published by James R. Hansen, a NASA historian, in 1995. According to Hansen, in the "lunar theaters" in Langley, every detail of the landing was tried out, just like a rehearsal in the theater, so that it was not possible to distinguish the real landing from a staged scenario. In fact, it only took a "small step" to pretend that a film made for simulation would become a big step for humanity with the landing. During this period, the New York Times wrote that "for those who had lost the race to the moon, there would be nothing but ruin and damnation", so to speak.

During subsequent Apollo missions, Von Braun was increasingly pushed away from what had been his dream. The scientist retired on May 26, 1972, a few months before Apollo's last trip, number 17.

7. Instead, we went!

Immediately after the publication of the shocking book by Kaysing, those who considered it blasphemy to doubt NASA and... television unleashed their response. The first debunkers, known as those who want to debunk hypothetical conspiracies, were the men of NASA, but not only them. Among the most seasoned debunkers are the Swiss Italian journalist Paolo Attivissimo, Phil Plait, animator of the site Bad Astronomy, Jim Oberg, an expert on space issues, Jay Windly, of the site 'Moon base Clavius,' and the duo Adam Savage - Jamie Hyneman, of the Discovery Channel program Mythbusters, which was broadcast from 2003 to 2016.

In particular, as Mazzucco reveals in his film "American Moon," Attivissimo is a real NASA collaborator, being the site manager of the 'Apollo lunar surface journal.' All of them, in one way or another, report a series of tests that they consider overwhelming evidence to prove that the moon landing conspiracy does not exist, with more prose or very prosaic prose.

The number one test of the landing would be Soviet silence. The USSR had undertaken a real space race with the US, and in fact, the Soviets were the first to send a human artifact to the moon since 1959 (Luna 2), repeating the feat several times to send the first images televised by the satellite in 1966 (Luna 9). How is it possible, therefore, that they did not know if the US moon landing was a staging and, knowing it, why did they not report the scam to the world?

Explanation number one: there would be a logical reason. Since the direct competitors of the United States in the space race (to send a man to the Moon, as we have seen, was not in their programs), everyone would have thought of a Soviet strategy of disinformation to throw discredit on the adversaries, and nobody in the world would have believed such a thesis coming precisely from the Kremlin. Half a billion human beings had seen the moon landing "live" on TV. A Tass communiqué in the opposite direction would have done nothing but backfire against the USSR. Moreover, the Soviets certainly did not agree to report a possible forgery. In fact, during the Apollo missions in progress, US-USSR negotiations had already begun for spatial collaboration, which would then lead to the joint realization of the International Space Station. The Nixon presidency, known as the president of the Moon, was pursuing a policy of thawing with the entire communist bloc, starting with the Soviet Union and then China. Nixon himself appeared on TV to give importance to this aspect of his foreign policy, "so that the two most powerful nations in the world can live together in cooperation." On May 24th, 1972, less than a month after the conclusion of the Apollo 16 mission, the penultimate program, Nixon signed a historic space cooperation agreement with the Soviet Union in Moscow. Less than a year after the conclusion of the Apollo program (December 19, 1972), American astronauts went to the USSR to meet the Soviet cosmonauts, participating in joint press conferences, visiting the

Baikonur space center, simulating space activities within the Soyuz simulators, and paying homage to the graves of the great Soviet pioneers of the conquest of space, first of all, Gagarin. After a few months, the Soviet cosmonauts returned the visit to the Americans, going to see the Houston space center and the Apollo shuttles.

Astronauts and cosmonauts, as well as all the technicians from the two superpowers, were busy setting up a system of coupling in space between the two types of spacecraft. This fervor reached its climax on July 17th, 1975, when Soyuz 19 and Apollo 18 met in orbit, engaging each other and allowing three astronauts and four cosmonauts to visit each other in space. This mission was the first step towards the creation of the International Space Station. Therefore, if the Soviets had falsely claimed the moon landing, not only would nobody in the world take them seriously, but they would have also compromised the most important spatial collaboration in history that was being developed.

The second argument made by the debunkers is as follows: if over 400,000 people have contributed to the Apollo project in one way or another for more than ten years, how is it possible to buy the silence of all these people over 40 years? In practice, this is a larger population than that of Florence.

Explanation number two: In reality, all these hundreds of thousands of people worked for a myriad of private companies scattered throughout the United States, with each company in charge of small parts of the project and working independently from the others. None of them would have been informed about a plan to falsify the landing.

Why is this the case? As Mazzucco recalls in 'American Moon', when a soccer player or boxer decides to sell a match, only he and perhaps his coach (in the case of the boxer) know what is happening on the field or in the ring. Another striking example is the 2015 Volkswagen 'Diesel gate.' The German automaker has over 600,000 employees worldwide, and certainly not everyone was aware of the scam that Volkswagen was putting into effect in the United States regarding the emissions of its diesel cars. A small command group was enough to manage this deception, including their employees. In the Apollo case, we saw that Bob Gilruth's Space Task Group managed everything, which consisted of less than 80 people, all bound by military secrecy, including the astronauts. Only they knew what was happening in orbit and during the video transmissions broadcast to the world. The official story tells us that the images of Armstrong and Aldrin's first lunar walk were broadcast live from the moon. However, if the same sequence had been prerecorded in a studio and transmitted back to Earth via satellite, nobody would have known except for the technicians at the Houston control center and those of us who watched it live on our black and white televisions.

• The third piece of evidence in favor of the moon landing is the 382 kilograms of "lunar" stones brought back to Earth by the Apollo missions. How could they not come from the Moon?

Explanation number three: How could they come from somewhere else? The official thesis states that the rocks collected from the lunar surface are much older than the rocks on Earth. Using radiometric dating, the estimated age ranges from 3.16 billion years for basaltic samples from the lunar seas to 4.5 billion years for those from the continents. In contrast, the oldest rocks on Earth date back to 3.8 billion years ago. Although it is assumed that their enormous age is verifiable with such precision and their composition is different from that of terrestrial rocks, it has been known for many years that there are numerous meteorites of lunar origin discovered in Antarctica. This fact can become a further clue; a mysterious journey made in Antarctica by the father of rockets Saturn, the Nazi scientist Werner Von Braun. In January 1967, just over two years after the Apollo 11 mission, he left for the South Pole, officially for a "personal vacation." Von Braun was hosted in an American military base, the McMurdo Station, in an area where a mountain was even named after the German scientist, Mount Von Braun. On this "vacation," Von Braun was not alone with his wife or friends but brought with him Maxime A. Faget (from the NASA Space Task Group), Robert Gilruth (director of the NASA Manned Spacecraft Center), Ernst Stuhlinger (another German rocket scientist working in Germany with von Braun), and two other unidentified scholars of the so-called Deep Freeze Project, a series of Antarctic exploratory missions carried out by the United States of America beginning in 1955-56. They were supposed to be two geologists.

The New World Encyclopedia online reports that Von Braun's 'was one of the first missions to fathom the icy surface in search of lunar meteorites to be used later as touchstones.' Who can ever know if NASA has presented these stones found in

Antarctica as lunar rocks actually taken on the spot? Who could have prevented such an easy forgery? Moreover, returning to the composition of the rocks, some doubt that they are lunar. Bill Wood, for example, a scientist who lives in California with degrees in Mathematics, Physics, and Chemistry. Wood has worked with many companies contracted by the US government for national security projects, including McDonnel Douglas. For many years, Wood has doubted the authenticity of the Apollo missions. "All the thousands of professionals who have collaborated on the project want to believe in it at all costs because it is their pride and their life. They cannot admit that this is a scam and besides, all of us Americans are very proud of such an achievement," he says in a famous interview. "As for the hundreds of kilos of moon rocks we claim to own, I think they were produced on earth. They may have been irradiated or placed in a vacuum environment, however modified in some way to make them look different."

An episode that caused a stir was the alleged lunar rock donated by the US in '69 to the Rijksmuseum in Amsterdam. As it was revealed 40 years later, in 2009, it is actually a piece of petrified wood, but probably, in this case, it is a scam fraud, since it would only be an artistic performance in which many have taken the bait, and they present it today as evidence against the lunar scam.

• Test number 4 of the real moon landing would be the panoramic views (even at 360 degrees) of lunar landscapes, which would have made it impossible to create them in a movie studio.

Explanation number four: This thesis demonstrates a lack of knowledge about the possibilities of cinematographic creation. We are not just talking about the special effects discovered and used by George Melies in the early 1900s. In 1915, Dwight W. Griffith's masterpiece "Birth of a Nation" had already simulated immense panoramas with tens of thousands of extras, when in reality there were only a few hundred. More recent productions, such as the 2005 documentary "Magnificent Desolation" based on the book of the same name by Buzz Aldrin, and the previously mentioned series of films, show that the evidence presented by debunkers is inconsistent. The filming takes place in a relatively small studio, and the final result is achieved through special effects that progress with technological evolution but are essentially based on the addition of painted, projected, or virtual images. This creates the illusion of an extremely deep or extended landscape. During the Apollo 11 years, the so-called "front projection" technique was used, and one of the first directors to use it was Stanley Kubrick for "2001: A Space Odyssey."

Industry experts can recognize a cinematographic image with front projection at first sight because the real image in the foreground is separated from the projected images by an imaginary continuous line throughout the frame. Unfortunately for those who believe that the Apollo missions were real, a line of this kind exists in almost all the images of the Moon spread by NASA, including those at 360 degrees. Another clue that suggests that the master of British cinema may have been involved in NASA's fraudulent project is the almost unknown name of

Frederick Ira Ordway III. Frederick, a Harvard graduate, was a space enthusiast and a member of the American Rocket Society since 1939. He was a friend of Werner Von Braun and worked at the Marshall Space Flight Center in Huntsville, Alabama, where the Saturn rocket was developed until 1963. Above all, Fred was simultaneously hired as a scientific supervisor by both Kubrick for the movie "2001 Odyssey in Space" and by NASA for the Apollo project. Is this a coincidence?

However, the technique of front projection alone would not be enough to be convincing in a possible lunar fraud. Hollywood workers are also specialized in the construction of models that are so perfect that they are almost indistinguishable from a real landscape. The most striking example is the futuristic city built for Ridley Scott's masterpiece, "Blade Runner," in 1982. In a model of just over 5 meters by 4, we see on the screen, at the beginning of the film, a veritable megalopolis. The Lem, the astronauts' return form, could also be included in the model category. In fact, when the Lem restarts from the dusty lunar soil (as shown by Aldrin's footprints), it does not raise any cloud of dust. Later on, there are also strong doubts about its "rocket engine."

Explanation number five: The fifth proof that the moon landings have taken place would be the images transmitted by various probes, such as the Japanese Kàguya, which recently detected traces of all Apollo missions and mapped the surface of the moon up to 15 kilometers of altitude, allowing for 3D processing. However, this thesis proves nothing because since 1966, NASA had sent five probes to explore the moon's surface in detail, the Lunar Orbiter, celebrated in several propaganda documentaries of the American body. These probes did not go any closer than 40 kilometers from the surface, but this was enough to map the Moon to 99 percent, with a maximum resolution of 1 meter per pixel. The total number of high-resolution photos was 2,180, and there were 882 in medium resolution, as NASA itself had stated. Thanks to these photos, NASA was able to build a huge lunar globe or model. It is worth considering how much effort it would take to retrieve the model from a camera and simulate it as if it were the Moon. Considering that NASA had the foresight to have a nice circular rail built around the model, doubts can only be fed. If a camera with a macro lens had been used to make the model almost touch the lens, the entire landing phase would have been more than likely, transmitting the images on the screen of a simulator as if they were really coming from the Moon. Moreover, the NASA lunar simulator really existed since 1964, and it was proudly shown in propaganda documentaries. Officially, it was used to train astronauts, but who would prevent NASA from using it for other purposes?

8.The deadly Van Allen belts

Ironically, it was NASA's probes that discovered the existence of two bands of plasma emitting high-power radiation in the Earth's magnetosphere, which is retained by the magnetic field of our planet. These bands were discovered in 1958 by physicist James Van Allen, who defined them as "radioactive belts." Although a third belt has apparently been discovered lately, the Van Allen belts are commonly known as two. The first belt, according to spatial measurements, extends between 1000 and 6000 kilometers from the Earth's surface, while the second belt goes from 10,000 to 65,000 kilometers (the Moon is about 400,000 kilometers from Earth) and is particularly active and insidious between 14500 and 19000 kilometers of altitude.

The Van Allen belts are therefore among the main sources of radiation in the space closest to us, causing electronic disturbances aboard satellites and probes and rapid degradation of solar panels. These radiations are mainly composed of electrons, protons, neutrinos, positrons, photons, and other particles with ionizing power that can break the covalent bonds between atoms. The most serious problem that can happen to an electronic device passing through the belts is the so-called 'Single Event Effect' or SEE, a phenomenon caused by highly ionizing particles that can cause immediate malfunctions of one or more transistors capable of affecting the whole circuit.

The "Clementine" probe, launched on January 24th, 1994, as part of the Deep Space Program Science

Experiment (DSPSE) space program, provided a concrete idea of the damage that could be caused by the Van Allen belts. It aimed to achieve scientific observations of the Moon and some asteroids. Almost four months later, on May 7, 1994, the central computer sent an unintended command that caused an altitude controller to burn all the fuel, resulting in a failed mission. This incident occurred despite the technology being much more advanced than that of the Apollo missions and the electronic components being extremely resistant.

It is not a coincidence that so far, all space missions with men on board, such as Soyuz, Gemini, the Shuttle, and the International Space Station, have been held far below the dangerous radiation zone, rotating in what is called 'low Earth orbit.' The only living things in the history of the universe to have crossed the Van Allen belts would be the 18 astronauts sent to the Moon.

Van Allen himself wrote in 1959 in the scientific journal 'Scientific American': "The discovery can be a big problem for astronauts. Somehow it will be necessary to adequately shield their bodies from this high-intensity radiation, even if they pass through this region very quickly."

Still, in 1961, Van Allen published another paper in Space World magazine entitled "The Dangerous Area: The Earth is Shrouded in Two Deadly Radioactive Belts." With regard to the inner band, he added, "Because of the great penetrability of high-energy protons, effective shielding is far from our engineering capabilities in the near future." Peremptorily, Van Allen's article concluded, "All attempts at human flight into space must be kept away from these two bands of radiation until adequate protection for astronauts is developed." It is no coincidence that NASA was very active in studying the effects of human tolerance to radiation. In those years, a plastic mannequin was developed in the laboratory of the military base in Kirtland, New Mexico, which, according to official announcements, had the same atomic composition as human tissue. This was done to verify the depth of penetration of the radiations inside the body.

As we have seen, the radiation problem was taken very seriously. However, after the shooting of Kennedy, when NASA was commissioned to organize space missions to send men to the Moon, the problem of the Van Allen belts disappeared inexplicably from the radar. Not even an obvious attempt was made to send an animal first to cross the radiation belts, as the Soviets had done before sending Gagarin into orbit around the Earth. So, on December 21, 1968, Apollo 8 started from the Kennedy Space Center in Florida with a lunar orbit destination. But even on this mission, there were many who raised doubts, not only about the Van Allen belts.

As reported by the website legamedelcielo.it, registered anonymously in 2003, just a few calculations can be made to deny NASA's official arguments about the Apollo 8 trip. The following information was reported on the website:

Departure on December 21, 1968, at 12:51 (UTC)
Return on December 27, 1968, at 15:51 (UTC)
Declared 10 lunar orbits at about 181 km (perihelion) and 191 km (aphelion) from the surface

The first "strange" element is the time it took to make the lunar orbits, because if in this first mission they declared that 30 minutes were enough to pass on the back of the Moon, and therefore that a complete turn could be done in about an hour, maybe their calculations in this first mission around the Moon were incorrect and then corrected. In fact, it is easy to make an approximate hypothesis of the time it takes to circumnavigate the Moon:
If the speed around the Moon was about 5,000 km/h, and the circumference at about 110 km from the lunar surface is (2πr = [2 x 3.14 x (1738 + 110)] =) 11.605 km, at such a speed it would take about 2 hours and 20 minutes to circumnavigate (t = S / v = 11605/5000 = 2h and 20m). For other speeds, we have:
- 6,000 km/h = 1 hour and 56 minutes.
- 7,000 km/h = 1 hour and 39 minutes.
- 8,000 km/h = 1 hour and 27 minutes.
- 9,000 km/h = 1 hour and 17 minutes.

To understand the speed at which the Apollo astronauts went during lunar orbits, we can look at the Apollo 11 mission, in which they performed 30 lunar orbits in a total time of 59 hours and 30 minutes. Therefore, for that mission, a tour was completed in about 2 hours (1 hour, 59 minutes). Regarding the claim of fakery, it is suggested that the total time of the mission was 6 days and 3 hours, including 10 lunar orbits that resulted in a 20-hour waste of time (not 10 hours, as stated by NASA, declaring a radio silence of 30'). This means that for the outward and return journeys it took around 63 hours and 30 minutes on average. This time is far below that declared in the other missions, where the calculation was well over 72 hours. With a single orbit and the time of descent, it was over 100 hours (more than 175 for Apollo 11).

In addition, the same source that claims fakery also questions the photos taken by Apollo 8. The lunar phase of departure needs to be verified to compare the images of the mission with the reality of the position of the Moon and Earth. In the photos proposed by the site, there are projections of the image of the Earth and the Moon calculated with the "Stellarium" program. Setting the date of December 21, 1968, and the position of the observer for the Moon in its center and for the Earth in the south of Italy, it is found that on that date, the Moon was the first day after the new moon, so the part visible to us was almost completely in darkness. Three days later, on the occasion of the "lunar orbits," only a quarter of the lunar surface visible to us was illuminated.

The source questions how filming or photos of the area of the future landing (the Sea of Tranquility) could have been made if that site was still completely in the dark. The few photos taken by NASA related to that mission are so small that they cannot prove anything. The first 83 photos - out of a total of 104 (for a mission of this importance?!) - are of absolutely useless land elements. Only the remaining 21 (less than a roll of 24!) are in space. If there was indeed a space trip, with the exorbitant costs of an Apollo mission, how is it possible that the three astronauts, Borman, Lovell, and Anders, took only 21 images? Is it not that by chance all their photos were then used for future missions?

For a direct view of the inconsistencies in the photos, visit the site legamedelcielo indicated in the note.

Returning to cosmic radiation, in his film, Mazzucco points out that the Apollo 8 spacecraft was protected only by a thin coating of aluminum, yet the radiation risk never seems to have worried the astronauts during their radio conversations with ground control. In the final report of the Apollo 8 mission, a 250-page document, the crossing of the Van Allen belts is never even mentioned. Also, for all the following 8 lunar missions, NASA has always minimized the amount of radiation absorbed by the astronauts in the double crossing of the bands, the outward and the return. For Apollo 11, an absorption of 0.11 rad was declared, and for Apollo 14, at the highest altitude, 1.14 rad was declared. Considering that a dose less than 100 rad/hour does not have any effect except some changes to the blood cells and that a dose is considered deadly when it exceeds 300 rads/hour, the amount absorbed by the astronauts made them laugh.

Unfortunately for NASA, even in this case, there are those who think differently. Assuming that the speed of exit from the Earth's orbit of the Apollo spacecraft was (as officially communicated) 25,000 kilometers per hour, the time of exposure of the astronauts to the protons of the first band and to the electrons of the second band would have been 52.8 minutes for each mission. The calculation leads to an absorption of at least 11.4 rads each time, despite the fact that any spacecraft could have been protected.

But then why, in 2014, 45 years after the 'lunar conquest,' did the NASA engineer Kelly Smith, in charge of the guidance and navigation systems of the future shuttles of the Orion program, which should bring humans back to the moon in 2030, declare: 'Moving further from the Earth, we will pass through the Van Allen belt, an area charged with dangerous radiation. Radiation of this type can damage driving systems, onboard computers, or other electronic components. [...] The sensors on board will record radiation levels for scientists to study them. Should we solve these problems before we can send people beyond this region of space?' Before'? Either the engineer is totally helpless or unintentionally made the revelation of the century. Twenty years earlier, in 1993, NASA's Space Radiation Analysis Group (SRAG) made an official statement referring to Apollo missions and solar activity: 'Among the Apollo 16 and 17 missions, there was one of the most intense solar flares, which produced radiation levels such as to represent a lethal dose for astronauts who, outside the magnetosphere, had been exposed to it within 10 hours of blasting. The fact that the timing of this blasting did not coincide with the Apollo missions is very fortunate. The accurate prediction of these solar events is not yet possible, in fact. Furthermore, the biological effects of long-term exposure to high levels of cosmic radiation are not yet fully understood.' And, again, to quote the statement of Ellen Stofan, former NASA chief scientist and adviser to NASA's Chief Executive Officer up to 207, Charles Bolden: 'NASA's goal now is to send humans beyond Earth's orbit to Mars. We are trying to develop new technologies to get there. At the moment, it is a technological challenge, not just because there are a couple of relevant issues. First of all, radiation: once out of Earth's magnetic field, astronauts are exposed not only to solar radiation but also to cosmic radiation. These are higher doses than we think a human being can absorb."

Stofan is not the only person who believes that radiation is still a crucial problem today. Robert Naeye, a former NASA contributor and senior contributor to "Sky & Telescope," an online monthly publication about astronomy, stated in an interview on NASA's website on August 23, 2007:

"Surprisingly, the Moon itself is a source of gamma rays. Its surface is heavily exposed to cosmic rays and solar flares. When cosmic rays hit the surface, they produce a deadly vapor of secondary particles right at the feet of those who walk there, giving rise to small nuclear reactions that release many forms of radiation in the form of neutrons. It is the same surface of the Moon that is radioactive!"

Sadly, the topic of the Van Allen belts may end with the disconcerting interview of Alan Bean, who was already aboard Apollo 12. When asked if he had encountered any problems during the crossing of the radiation belts, Bean replied, "No, I'm not sure if we were far enough away to cross the Van Allen belts. Maybe we did." Only when he was told that the belts are located between the Earth and the Moon did Bean say, "Then we certainly passed through them!"

9. The Lem and other pieces of junk (like the Lunar Rover)

The Lem, also known as the Lunar Excursion Module, was the descent module that the astronauts used to set foot on the Moon, and it was a remarkable achievement of engineering. The project and construction of the LEM were carried out by the aerospace company Grumman for NASA between 1962 and 1969. The compensation for Grumman's tender was set at $385 million at that time, including a net gain of $25 million for the company, despite the fact that they only had a vague idea of the machine that had to be built. As a result, the project started a year later than the other members of the Apollo program.

The design of the LEM began shortly after Kennedy's famous statement on the conquest of the Moon. At that time, the machine seemed so unlikely that it was nicknamed the "flying canopy." The problem that seemed almost insurmountable was how to make the module stable during the descent since it was not possible in space to rely on any aerodynamic support.

Just a year before the landing, Neil Armstrong risked his life because of a malfunction in the Lem. At 70 meters height, the Lem became ungovernable, and there was nothing left for the astronaut but to eject with a parachute before the Lem crashed to the ground. Three of the five Lem test specimens also crashed during the tests, but the 16 engines they were equipped with made them much safer than the final version of the Lem.

In the end, the changes to the initial design were so numerous and contradictory that they resulted in the twisted shapes of the cabin. In some cases,

Grumman's engineers chose to assemble the elements of the Lem structure with rivets instead of welds, despite the great opposition of the NASA managers who doubted the stability of the pressurized cabin.

However, this is still nothing compared to the fundamental importance of the LEM as a survival form for the two astronauts who lived there during the descent and the departure from the Moon. Temperatures on our satellite are extreme, with no atmosphere to regulate them. When the sun is shining on the Moon, temperatures can rise to 130 degrees Celsius, while in the shadows, temperatures can drop to less than 100 degrees Celsius. The Lem would have faced the challenging situation of having one half scorching hot and the other half icy cold, like the most remote arctic ice. It is difficult to imagine what kind of structure could have been conceived with the technology and materials available in the 1960s that could sustain such extreme temperature differences. Additionally, how could we ensure that the astronauts could pilot such a delicate module, with fuel strictly counted, and with safety parameters almost at zero? On board, there was a computer developed by the Massachusetts Institute of Technology that used integrated circuits, a groundbreaking technology at that time. However, this feature also made it extremely unreliable since there was no effective testing for this technology in the early days. Remote driving from Houston had been excluded due to the latency of radio transmissions between the Earth and the Moon of more than three seconds, which would have prevented an immediate reaction in case of problems. For all these reasons, many people believe that given the technical difficulties involved in not only the structure but also Lem's control system, it could have been nothing more than a model to be shown on TV.

Looking closely at the images provided by NASA, it seems difficult not to accept this theory. While the Lem looks like a real landing shuttle in the totals, upon magnifying the images, you can see steel tubes, disconnected aluminum panels, rivets placed at random, engines covered with tinfoil paper, real sheets of cardboard, and even scotch tape used in the construction. It is hard to imagine how a structure of this kind could withstand the extreme thermal shock.

One of the claims made by NASA technicians was that during the outward journey, they had to give a slow rotation to the shuttle to avoid overheating the delicate equipment of the Lem, as it was built to function exclusively in space. However, it is puzzling how it could easily withstand the rays of the sun for almost three consecutive days on the moon.

In spite of Grumman's already high budget, the Lem ultimately cost a staggering 2 billion and 200 million dollars of the time, which is equivalent to over 21 billion dollars today, practically the cost of an Italian financial institution. But the concerns about the Lem do not end there.

Once the module was "lined up," no crater was visible beneath it in the lunar sand that would have been created by the engine's jet during descent. According to "believers," the crater does not exist because the Moon's gravity is one-sixth that of Earth, and the Lem used low power to land. Furthermore, the Moon's surface is rocky, akin to volcanic terrain, as described by Attivissimo. However, the astronauts denied this thesis, with Neil Armstrong stating in a radio transmission that the surface seems to be made of very fine grains, almost like dust. Moreover, Michael Collins's boot imprint on the Moon's "sand" is visible to all.

A third suspicious problem concerning the Lem is the absence of flames under the rocket engine during the restart from the Moon. The module consisted of two separate blocks, one lower and one upper, with the descent stage serving as a launching pad for the ascent stage destined to return the astronauts to the Apollo shuttle. Both stages had their own independent engine, with the lower one for landing and the upper one for restarting. Both engines were "hypergolic," meaning they burned spontaneously when combustion and fuel were brought into contact, greatly reducing the possibility of ignition faults.

According to those who believe in the landing, the lack of flame under the Lem during restart is due precisely to this type of fuel. However, this claim is denied by dozens of films that show how hypergolic engines undoubtedly develop flames both on Earth and in the absence of an atmosphere. Other doubts also arise concerning the restart of Lem, which will be explored further in the chapter on lunar dust. This substance, called regolith, is essentially a rocky material that can range from the composition of gravel to a few microns in diameter for each grain. Despite Grumman's already high budget, the Lunar Excursion Module (LEM) ended up costing a whopping $2.2 billion at the time, which is equivalent to over $21 billion today, almost as much as an Italian financial institution. But the concerns about the LEM don't end there.

Once the module was "lined up," no crater was visible beneath it in the lunar sand that would have been created by the engine's jet during descent. According to "believers," the crater doesn't exist because the Moon's gravity is one-sixth that of Earth, so the LEM used low power to land. But the astronauts denied this thesis. "The surface seems to be made of very fine grains, almost like dust," said Neil Armstrong in a radio transmission, while Michael Collins left an imprint of his boot on the "sand" of the Moon.

A third suspicious problem with the LEM concerns the absence of flames under the rocket engine during the restart from the Moon. The module consisted of two separate blocks, one lower and one upper. The descent stage allowed the astronauts to land on the Moon and served as a launching pad for the ascent stage, which returned them to the Apollo shuttle. Each stage had its own independent engine, with the lower one for landing and the upper one for takeoff. Both engines used "hypergolic" fuel, which burned spontaneously when the combustion and fuel were brought into contact, greatly reducing the possibility of ignition faults. Believers claim that the lack of flames during the restart is due to the type of fuel used, but this claim is denied by dozens of films that show how hypergolic engines develop flames on Earth and in the absence of an atmosphere. Skeptics also doubt the restart of the LEM, as regolith, a rocky substance that makes up the lunar dust, can create hundreds of microscopic impact craters between 30 and 60 micrometers in diameter caused by particles traveling between 400 and 1000 meters per second. According to physicist Phil Metzger of the Kennedy Space Center, "If there are no mountains to obstruct the route, the sandstone raised by the exhaust could dart around the Moon" and land again at the foot of the rocket.

Moreover, particles of this size can creep into small cracks and crevices of any human-made machine, including the many probes already present on the Moon at the time of the Apollo 11 mission, posing a threat to subsequent missions. It seems unlikely that the designers of the LEM never thought about this concrete threat to the safety of the astronauts, especially with a structure so light and well-assembled. Was it an animated model? If Kubrick and even Disney had managed to make them credible, it's unclear why NASA and its designers couldn't have done the same.

Why not take a look at the original projects of the LEM to see what a technological marvel it was able to land and leave on the Moon in 1969? Today, it would be possible to simulate its actual effectiveness using modern computers. Wouldn't that be great? Unfortunately, as in many other cases, NASA is no longer able to provide the original projects of the module, as the construction company Grumman threw them away because they "occupied too much space." Is it logical to throw away the project of Galileo's first telescope or the first internal combustion engine? Unfortunately, in the case of the first machine that allowed a human being to arrive on another celestial body, that's precisely what happened. Has Grumman never considered donating these projects to a museum? "To think badly is a sin," said Andreotti, "but almost always there is something right."

It may seem impossible, but there is something even more disconcerting than throwing Lem's projects into a dumpster: the disappearance of the original tapes that recorded Armstrong and Collins' first lunar walk. And why did NASA admit to their disappearance only 40 years after the 1969 mission?

The most important document in the history of humanity, which certifies the presence of a human being on another celestial body, "is no longer found". This is so naive that it wouldn't even happen in an Agatha Christie mystery.

Former Apollo 11 flight director Gene Krantz, when interviewed, said, "I have found no evidence that the tapes with telemetry data exist. Even if they existed, we wouldn't have the right machines to read them today. At the moment, I can say that the telemetric data of the mission has disappeared." David Williams, one of the NASA archivists, also declared,

"We are not able to trace them. We have no idea where these telemetry data ended up or what path they could have taken in these years."

To justify this enormous lack of certain evidence of the moon landing, the debunkers, led by Attivissimo, offer this disconsolate explanation: "NASA has lost them for a purely bureaucratic reason. At the time these tapes were considered unusable. The TV signal was recorded on a track of the telemetry data, i.e., the pulse, the status of the vehicle, etc. According to practice, after a certain number of years, if no one asks for the data anymore, the tapes are sent for deletion because the tapes cost and do not leave them there to rot..."

I want to emphasize Attivissimo's last words, because it is thanks to statements like this that the theory of lunar fraud can increasingly spread. What kind of explanation is this? It is like saying that the first book printed by Gutenberg, since nobody asks to read it anymore, it will be good to send it to the pulp because the paper is expensive.

The loss in the case of Apollo 11 is not so much the images of the walk, of which there are other copies, but the telemetric data that also included the actual data on the positions of the Apollo and the Lem in space. If the situation in space had been different from that universally disclosed with the lunar history, it could easily have been verified or even easily confirmed from that data. Mazzucco in American Moon then asks, "What was so important about those tapes so as not to make them more available for verification by inventing an excuse as trivial and silly as that of loss?"

Apart from the elusive Lem, starting from the Apollo 15 mission, another technological 'barrack' makes its appearance: the Lunar Rover.

The first to conceive the idea of a car usable on the Moon was Werner Von Braun, who began publishing a series of articles in Collier's Weekly magazine entitled "Man Will Conquer Space Soon!" in 1952. Von Braun imagined a Rover suitable for a six-week stay on the Moon with the ability to move weights up to 10 tons to build a space base. Not surprisingly, shortly thereafter, in 1954, he was hired by Walt Disney as Technical Director of Disney Studios, and he also worked on the 1955 documentary "Men in Space."

In just six feet of the last stage of the Apollo missions, NASA had managed to get astronauts, computers, equipment, engines, and, of course, the strange object called Lem.

Lem, which during the missions from 11 to 17, never changed its constructive characteristics, yet starting from Apollo 15, another small miracle took place.

In the belly of the Lem, technicians from various important American construction companies managed to bring in a Lunar Rover. However, with its length of about 3 meters and 10, and width of 1.14 meters, getting it in was a challenge. The lower part of the Lem was already filled with survival equipment and the landing engine. The ingenious solution studied at NASA was to fold it up like an origami.

There is only one video available on the supposed Luna that shows the extraction of the Rover from the Lem, and it is from the Apollo 15 mission. In the video, the Rover appears to unfold almost by itself, as if powerful springs had been inserted to make it stretch alone. The two astronauts (Scott and Irwin) eventually move the Rover, which appears very light due to the lack of seats, batteries, transceiver group, antenna, camera and data/video transmission equipment, all of which are necessary on Earth but not on the Moon. The whole small frame looks very different from the prototypes officially presented on Earth. It is unclear whether it was the same chassis. If the extraction scene had been filmed on Earth and not on the supposed

Luna, the Rover, which officially weighed 210 kilograms, would necessarily have to be much lighter than the real one that appears later, just to simulate its reduced weight on the Moon (35 kilograms).

Another question is how Scott and Irwin managed to assemble all those dozens of missing elements from the Rover while wearing their space suit gloves.

One of the most notable experiments of the Apollo 15 astronauts was the famous demonstration of Galileo's gravity. In 1971, astronaut David Scott demonstrated that a feather and a hammer could fall at the same time and at the same speed in the absence of atmosphere. The success of the experiment was taken for granted, but something does not seem right. To the naked eye, when the feather touches the lunar surface, it bounces. If it had been a true falcon feather, as stated by Scott, this would have been impossible. Was it perhaps the same golden feather that was later presented as a testimony left for the death of the many astronauts who had preceded them? Moreover, the handle of the hammer held by Scott seems to sway. Why? It should stick in the sand like a rock, but instead it seems to behave differently. Where did the sand of the Moon go in which the footprints of so many boots were imprinted?

10. Here Luna, to you Earth

One of the most perplexing aspects of lunar missions is the radio communication between the astronauts and the Houston space center. In all the Apollo films, even when the astronauts are on the Rover, radio communications never cease, and they are seamless and without any delay. Is it possible that, in 1969, radio signals could be transmitted and received from almost 400,000 kilometers away with a small antenna on a helmet to Earth? Furthermore, from Apollo 15 onwards, the astronauts weren't even connected by cable to the Rover antenna. Of course, an antenna could have been used on the Lunar Excursion Module (LEM), and they could have communicated through it, but this was never shown. Perhaps Houston could receive a faint signal from the Moon through enormous terrestrial parabolas, but with which antenna did the astronauts receive communications from Earth? An even more paradoxical aspect of these communications is the lack of delay between them. According to physical laws, given the Earth-Moon distance of 384,400 kilometers, a radio signal should take about 1.3 seconds to arrive and the same time to return, making a total of 2.6 seconds, a little over two and a half seconds. Yet, many conversations between Houston and the astronauts are punctuated by much shorter time spaces, sometimes with answers that are almost instantaneous. This is inexplicable from a physical point of view.

This is what the website legamedelcielo says about the radio communications of the Apollo missions intercepted by other listening centers and by many amateurs in those years: "If the men inside the Command Module never went to the Moon because they lacked sufficient thrust, it was necessary that something else certify to the free and collaborative scientists (Australians, Russians, Germans, English) that such a journey was actually taking place. The third stage of the Saturn V could have done it autonomously, reaching an inertial speed sufficient to reach the Moon, like any other mission for the setting of lunar orbit or landing on the lunar surface of satellites/robots previously made [...]. In fact, considering that the LEM was not necessary, the void of the rocket could contain/hide the necessary elements so that there was a transmitter for the outward journey up to its placing in lunar orbit and another for communications from the surface, containing the instruments that would later certify the landing in a declared point and ratify the Doppler effect as real. The third stage was abandoned in space or precipitated on the lunar soil. Meanwhile, the real astronauts remained silent and rotated into Earth orbit, awaiting the completion of the false mission. Since the false images in weightlessness could have been made before the actual Apollo missions, there was no need to build a real capsule and service module."

According to the anonymous author of legamedelcielo and many others, everything shown live during the moon landing was only pre-recorded videos shot in simulators and dubbed with the voices of real astronauts, made with photos already acquired by lunar probes, which were aired as if the events were taking place live, including false landings, sequences on fake lunar soil, restarting and hooking into orbit with animated models.

The real journey back, having nothing left to offer, would have proceeded in absolute silence.

Speaking of communications, let's now examine the fantastic 'live TV' from the Moon. The first three missions - the 11, the 12, and the 14 - had really poor video quality. Starting with Apollo 15, there was a technological breakthrough. With the brand new Lunar Rover, the astronauts could now also count on a camera mounted on their own jeep: the Rover TV. This camera was operated directly from Houston. The images were transmitted with an inverted umbrella antenna mounted on the Rover. The transmission of images could be done only to the immobile Rover and after an accurate pointing of the antenna. In this regard, the NASA website reads: "the high-gain antenna emitted a radius narrow enough to be intercepted on Earth by an 85-meter dish". The gain of the antenna was 24 dB at the center of it, reduced to 20.5 dB on a 10 ° cone. This means that the signal became weaker as the transmission angle increased. The pointing of the antenna was not a simple operation.

First, a careful manual alignment operation was required, followed by an even more accurate optical aiming, given that the Earth seen from the Moon subtends an arc of less than 2°. The antenna pointing had to remain - according to the NASA manuals - within 2.5° with respect to the Earth, and this was only possible when the Earth was in the center of the optical viewfinder with which the antenna was equipped. Beyond that point, the video signal would immediately become much worse. But the antenna was mounted on a Rover with wheels and suspension, so any undulating or jerky movement of the machine could affect the antenna's target. It was therefore essential that the Rover remained absolutely immobile during transmissions. However, in the original films, there are several cases in which the astronauts make panoramic shots or move the Rover, making it wince, but the transmission of images has no problems. The TV signal never goes off, nor degrades even a little, even if the astronauts get on and off the Rover during the shoot.

11. Moon Dust

The most iconic image of humans' supposed conquest of the moon is the footprints left by astronaut boots on the lunar soil, particularly the famous first ones left by Buzz Aldrin (Armstrong's footprints are not visible). The footprints are considered one of the safest "proofs" of human beings' arrival on our satellite.

As NASA later revealed, Aldrin's footprint was deliberately made so clear and distinct and then photographed for scientific purposes. It was necessary for researchers on Earth to "touch with their own hands" the compactness of the regolith, the lunar dust.

As can be seen in all the available photos, the impression is clear, perfect, and profound. This is incredible, considering that an average man weighing 80 kilos on Earth would weigh only 15 kilos on the moon while wearing a suit that according to NASA weighed 100 kilos. It would be difficult for such a person to make such a deep imprint on the ground, even if it were dusty. This is especially surprising given that the lunar dust does not seem as dusty when the Lunar Excursion Module (LEM) leaves the moon.

However, debunkers claim that the lunar chemistry in the vacuum is very different, and the stacking of grains on the moon, attracted downward by a force of gravity that is much weaker, has a lower tendency to collapse. Therefore, the edges of the imprints retain their shape more easily. Moreover, the lunar regolith has a remarkable electrostatic charge, causing the lunar grains to adhere to each other more than normal terrestrial sand does. This information comes from NASA, and no one can counter it unless they possess the lunar regolith. But what about the dozens of footprints left by the very light astronauts? How can they be as compact as if they were imprinted in mud instead of a dust that is totally devoid of humidity? Aside from the points already mentioned by Robert Naeye regarding the radioactivity of lunar soil, which causes "the vapor of secondary particles right on the feet of those who walk there, [to] give rise to small nuclear reactions that release many radiations in the form of neutrons," do you believe that the dozens of footprints left by the light-footed astronauts are as compact as if they were imprinted in mud, rather than in the dry lunar dust? On Earth, water is an essential element for shaping sand.

Those who blindly believe in the moon landing theory argue that various atmospheric agents on Earth constantly stir sand grains, making them almost spherical, whereas this does not occur on the Moon, resulting in lunar sand grains being angular and rough and more prone to getting stuck together. However, even recent attempts to replicate the impressions of astronauts in a vacuum room with a regulatory system provided by NASA have failed to produce the same clear forms as shown in the photos taken by the astronauts. Moreover, there is an even more puzzling aspect of the "lunar" dust: it behaves almost exactly like terrestrial dust. On Earth, dust tends to stick to objects due to atmospheric humidity, but on the Moon, there is no atmosphere, let alone humidity.

So, how can the astronauts' suits in the images be covered in dust almost up to their waists? "I am covered in dust from head to toe," says one, and "we are all dirty, you should see your back," says another. "You seem to have played in a coal tank," Houston remarks. Even the Hasselblad cameras themselves are not immune to the ubiquitous, sticky dust. "There is no way to keep the gnomon clean either," complains one astronaut. The gnomon is the instrument that acted as a chromatic reference for the photos. Shades of gray on the rod varied in reflectivity from 5% to 35%, and a color scale allowed for a more accurate determination of rock and soil colors by comparison.

According to the "believers," all this dust is simply due to an electrostatic effect, but this explanation is not plausible because the dust sticks to materials of completely different composition and shape, such as suits, lenses, mudguards, plastics, and so on. Moreover, NASA itself explains that the electrostatic effect on the Moon can occur only in shadows, and the solar wind would instantly cancel it.

In a document released in 2011, it is stated that "on the lunar surface exposed to the Sun where the photoelectric effect is predominant, the electrostatic charges dissipate almost instantly." Such doubts about the reality of the moon landing persist.

12. Ladies and gentlemen, the Circus!

Although each astronaut weighed an average of just 15 kilograms on the moon, it is unquestionable that there are films released by NASA in which astronauts perform feats with their bodies that cannot be explained by any physical law. They fall in all directions and get up without apparent effort, even in completely unnatural positions. This is unless we do not admit the existence of real cables that supported them. Rising up from the moon's surface, levitating is certainly not something that can be done easily with those cumbersome suits, even in the presence of one-sixth of the Earth's gravity. The problem of reduced gravity would certainly have been the main problem to solve in order to shoot the "conquest of the Moon" in a film studio. The technique used in these cases dates back to the time of the great Florentine architect Filippo Brunelleschi. He was the first to levitate a figurant in the role of the Archangel Gabriel along the nave of the church of San Felice in Piazza, in Florence. It was 1493, a year after the discovery of America. Since then, this scenic expedient has been used countless times in theater (think of the Momix or Cirque du Soleil shows), and cinema has also quickly taken it up, having an advantage over theater: in films, cables can be erased with tricks, photographic (yesterday) or digital (today).
Walt Disney made use of it in Mary Poppins, for example, just five years before the lunar missions, and it is currently used in all science fiction films. Today, cables can be completely eliminated in

digital post-production, but in the 1960s, the cinematographic procedure was much more complicated and did not always lead to perfect results. Every once in a while, some light reflected on a cable for a moment, revealing the trick to an experienced eye. This is what happens with some films identified by Mazzucco in his 'American Moon'. In the Apollo 17 mission, there is clearly something that reflects the light at least two meters above the astronaut's head, not attributable to the 30-centimeter antenna placed on his backpack. There are even sequences in which an astronaut rotates on himself suspended from the ground, just as if he were a ham hanging in the cellar. In another case, the astronaut slips with both feet forward and then backward without ever falling.

In short, if there is a reason to doubt the whole lunar affair, the mysterious and unconscious movements of the astronauts are certainly one of the most convincing. To understand the ease of creating such a collective illusion, you don't even need to bother Disney or Hollywood. A few years ago, in 2002, a London-based marketing company produced a video titled 'Moontruth', the truth of the Moon, in which the Armstrong scene descending from the ladder and (not) pronouncing the fateful phrase was replicated to perfection. Only 125,000 people have seen it so far, but if it were more widespread, it could certainly gain other followers of the 'lunar plot' theory.

Just at the time of the sentence "It's a small step for a man, a giant leap for mankind", in the background, a trellis that held a spotlight falls, the film crew enters the field, and the director shouts "Stop!".

"The lunar surface was cement dust. It was disgusting. Even with maximum studio ventilation, it would sneak everywhere, and at one point, it fluctuated so much that the lights produced an unwanted amount of reflections," said director Adam Stewart, who died on August 28 of the year following the video, at 31 years, due to a sudden, mysterious allergic food reaction. In the days following the publication of the video, NASA had received no less than 3,000 phone calls from people asking for the truth about the landing. And to think that the intent of Stewart and the marketing company was absolutely satirical. "We think our work is quite convincing, and one thing is certain: it cost much less than actually going to the Moon," they wrote in a note.

13. Greeting the flag

The Moon floats in the cosmic void and, like most celestial bodies we know, has no atmosphere. There is no air. However, NASA's films give rise to some doubts on this matter. Even if there is air, it certainly has not been turned on the Moon.

For instance, in NASA footage of the Lunar Rover on the Moon, small clouds of "lunar" dust raised by the wheels of the car can be seen, as if it were on any land beach. If there were truly no atmosphere, the sand raised by the wheels should make a perfect arc, very wide, before falling back to the ground. Due to Galileo's famous hammer and feather experiment, all particles, large and small, should fall together along the same trajectory. This is not the case on Earth, where the smallest and lightest particles are "lifted" or slowed down by the air in which they are thrown from the wheels, staying suspended longer than the heavier ones and falling even closer. All of this should not happen if we were truly on the Moon.

However, the most embarrassing evidence of the probable presence of an atmosphere in the films released by NASA is represented by some of the most iconic images of the Apollo missions: the American flags planted by all the crews on the 'lunar' ground.

In many films released by NASA, the flags seem to fly slightly, but in the vacuum, it is difficult to assess whether they are doing so only because an astronaut is touching their rod, or there is indeed some atmosphere. A solution to the dilemma seems to come with some videos presented by Mazzucco in "American Moon," in which a "lunar" flag can be clearly seen waving without any contact with the astronauts. In one case, that of Apollo 15, the flag sways as soon as an astronaut passes a few centimeters away from it, as if it caused a shift of air, which obviously cannot happen on the Moon. In another case, of Apollo 16, the two astronauts are photographed in front of the flag. Also in this case, the flag fluctuates slightly at the passage - without contact - of an astronaut in its vicinity. In the third case proposed by Mazzucco, the flag appears and disappears in a fixed frame without the astronauts even being in its vicinity (they even climbed up the Lem).

One can think of the "conquest of the Moon" as they want, but inconsistencies and suspicions always arise, regardless of the aspect of the missions that is being considered.

14. Impossible Photos

According to an old Anglo-Saxon proverb, "an image is worth a thousand words," suggesting that seeing something with one's own eyes gives it a value of reality. However, with the invention of photography, the manipulation of images has made it difficult to trust what we see. This cognitive relativism is even more apparent in cinema, which offers a window into previously impossible worlds. NASA's photographs of American astronauts on the moon during and after the "lunar conquest" have been scrutinized for years. Researchers from the USSR discovered that the moon was not gray as NASA's photographs suggested, but brown. Many inconsistencies in NASA's photographs have been noted, including the fact that the Hasselblad 500 EL/70 cameras used by astronauts were the same as those sold on Earth without any special protection from the Van Allen bands, solar rays, ultraviolet rays, gamma rays, cosmic rays, and lunar soil's radioactivity.

The films used by astronauts were normal terrestrial films that would have been compromised by the bombardment of atomic particles, resulting in a "granulation" effect on images. Moreover, even exposure to X-ray checks in airports could damage films, making it hard to believe that the films used by astronauts would have been perfect without any special protection. NASA documents have revealed that the annual flow of cosmic rays on the moon is 1.4279×10^8 particles per cm2 per year, making it impossible for the Hasselblads to function properly without protection.

NASA's inconsistencies in photographs and documents have led to doubts about the authenticity of the moon landing. The skeptics find it hard to trust the photographs and videos of the moon as they do not reflect the reality of the moon's environment.

From a document released by the agency in the late 1990s entitled 'Human Safety in the Lunar Environment', intended to examine the possibilities of human beings remaining on our satellite, we read: 'the annual flow of cosmic rays on the Moon, during the minimum of solar activity, is 1.4279 x 10^8 particles per cm2 per year, which means 4.5 particles per second per square centimeter.' A cosmic flood that penetrates any material, including the human body. The Hasselblads of the astronauts would have remained on the Moon even eight hours in a row without any particular protection. So, according to NASA, each roll would have absorbed nearly 130,000 cosmic particles, making them probably completely useless.

NASA is still shooting itself in the foot in another document dated September 1, 1971, in which, speaking of the Skylab project, the space laboratory that was put into orbit after the Apollo missions ended, it writes: 'They have been carefully studied the radiations present in the terrestrial orbital missions, and the results indicate that these can seriously damage the films, if not adequately protected. [...] The energy level of intergalactic cosmic rays is so intense that there is no practical method to avoid damage from cosmic radiation." The diffuse photos of the Moon, on the other hand, are perfect. Furthermore, we have already mentioned that the temperature of the Moon ranges from minus 100 to as much as 130 degrees centigrade. No film and no normal camera would ever be able to function in these conditions. Indeed, the films could crystallize or melt, depending on the extreme temperature.

But we are just at the beginning of the series of technical impossibilities. Examining any of the photos or videos of the Apollo missions, one cannot in fact do without thinking once again of Stanley Kubrick's masterpiece, "2001, A Space Odyssey". There are several clues that can make you think of a soundstage instead of a true lunar scenario:" Since two of the photos have been proven false, it is plausible that all three are false and therefore the similarity of the outlines can be explained: all three were taken on the same fake background. The change of scale in two of the photos by the professor is due to the fact that they were taken from a slightly greater distance than the other. The explanation for the different lengths of the external slopes is that these are adjustable features of the background. Suspicious slopes are present on both sides of the main ridge, which are rather undulating. To make the mountain look different from the three points from which it was "photographed", the sides were slid up or down, which effectively modified the apparent shape. The gift is the exact measurement of the upper profile. Once the veracity of these photos is questioned, many other questionable features become noticeable. All three photos have a clear "horizontal" line between the "flat" foreground and the background (front projection!). Such a horizontal line is a feature of many of the photos taken on Apollo's Moon. "My opinion," concludes Professor Rourke, "is that these photos were projected into the background by something like a large frontal projector. By carefully observing the distortion in the signs near the left slope while moving towards the upper ridge, a clear distortion is visible between the signs in the two figures and the signs approach a little closer to the edge in the S7 image. However, the distortion does not increase dramatically as you get to the edge, which should happen because this is not a normal edge, but is curved!" If you are not yet convinced of the possibility that NASA could have "invented" photos from the Moon, all you need to do is have a look at the long and fascinating study done in this regard by Jack D. White over the years. Jack, a Texan born in 1927 and a graduate in journalism, passed away in 2012, coincidentally when the Ukrainian professor Oleynik launched his justified accusations against NASA photos. Jack had been a photographer for half a century and throughout his life, he was a specialist in photographic analysis, to the extent that he was commissioned as a technical consultant by the Warren Commission investigating the assassination of JFK, the eponymous film by Oliver Stone on the Kennedy case. On the photos taken by the Apollo astronauts, Jack D. White found so many inconsistencies and anomalies that remain mostly unexplained.

Jack mainly focused on the reflections in photos, including those that captured the astronauts and lunar landscapes on the mirrored visors of their helmets, as well as "space" photos that depicted shuttles and the Earth in the distance. In his explanations, Jack showed countless times how NASA extensively manipulated the photos published on their site, adding or moving details as desired with the help of a photo editing program.

Of course, Jack took pleasure in finding discrepancies in every photo NASA provided for 50 years and, as a Photoshop expert, he was able to make unique discoveries in this regard. But he also wanted to warn us and transform some NASA scene photos into "authentic" lunar photos.

His message, which has also become his legacy, is clear today: "If I managed to deceive you, why couldn't NASA have done it?"

15. The Moral of the Moon Landing

In July 1970, one year after the Apollo mission landed on the moon, a newspaper survey revealed that 30% of Americans believed it was a forgery. Today, that percentage ranges from 6 to 9%, which amounts to 18-30 million people. In Great Britain, the percentage reached 25% in 2009, while in Germany, it was 46% in 2011. In schools in Cuba, Venezuela, Nicaragua, and Angola, the lunar conquest is taught as a staging.

In Italy, a recent YouGov poll revealed that 12% of Italians do not believe in the US moon landing, with another 10% declining to comment. Another survey conducted in 2017 by Moreno Mancosu, Salvatore Vassallo, and Cristiano Vezzoni raised the percentage of skeptics to 20%.

Personally, I do not intend to engage in a battle to promote the false landing theory over the official narrative. While I am convinced of my ideas, I deeply admire the will, effort, and dedication shown by hundreds of thousands of people, especially scientists and soldiers, who sacrificed their lives for over a decade to realize a dream that captivated an entire nation, even if stirred and dazzled by ambitious and troubled political figures.

NASA did an incredible job, whether the astronauts actually arrived on the moon or whether they were space puppets. The American organization could not afford to fail the mission after spending $135 billion from the US Treasury. Imagine what would have happened if average Americans had seen their taxes increase and social services diminish to serve Nixon's political ambition but without achieving any "concrete" results.

In 1971, almost at the height of the Apollo mission, President Nixon, also known as "Tricky Dicky," strongly promoted the birth of HMO (health maintenance organization), which led to the definitive collapse of the already precarious American health system, greatly reducing the costs for both the insurance company and the one who used the service. However, the truth was that they always paid less at lower costs. Can it be said that this is also the cost Americans paid for the frenzied rush to the Moon? While Nixon gloated beside the astronauts returning from the moon, he increasingly devalued public health facilities, which over time became crumbling and inefficient. In this way, most of the state health care was practically donated to the insurance companies that now manage the health sector in America. This reminds us of something? To learn more, contact Michael Moore and his 2007 documentary, "Sicko."

After arriving on the moon, terrestrial technology undoubtedly made giant strides without the need for war (even though there was a war). This was thanks to the bottomless pit that NASA could draw on. Our current science fiction imagery owes almost everything to those visionaries and opulent "prophets of space." However, it does not mean that we must accept every fairy tale told to us. Perhaps it worked with our fathers and grandparents, who were naive about communication in the 1960s and 1970s, but today we should be much savvier, especially when it comes to 'fake news.'

In addition to the small command group that managed direct communications and photos of the landing, it was necessary to have actors convinced of their part - the astronauts. As we have already mentioned, they were all career soldiers who had often fought in more than one real war. In such cases, it is not unlikely to obtain blind obedience. If there was also national pride, the flag, and the need to prevail over a "red" enemy, their collaboration at any level could not be questioned.

The first men to land on the Moon, Armstrong,
Aldrin, and Collins, splashed down in the Pacific
Ocean on July 24, 1969. An unparalleled triumph
awaited them, and according to the national
narrative, they had accomplished the greatest
human enterprise of all time.
After the inevitable media spectacle with Nixon, they
were held in quarantine for a few weeks and were
officially presented to the world press on September
16, almost two months after the venture. Looking at
the faces of the three "heroes" in Mazzucco's
documentary, "American Moon," it is hard to say
that they were the most popular men in the world.
To paraphrase John Lennon, astronauts at that time
were "more popular than Jesus."
When someone from the big press room asked what
the meaning of such an immense enterprise was,
the three looked at each other perplexed, and no
one really wanted to answer. It was not a military
question, after all.
The second pass awkwardly. "The mission
represents the beginning of a new era," Armstrong
finally whispered, looking first into the void, then
low, as if ashamed of those words just uttered. It
was a beautiful phrase to say to the press, just like
the "small step for man..." line.
Another disappointment was the question of how
the carpet of stars looked from the Moon. Well,
none of the three remembered having seen a single
star!
There was a stark difference with the three-mission
astronaut Shuttle, James Reilly, who remained in
space for over 35 hours. "You can see many, many
bright spots. Millions of points. And it's not like on
Earth because they don't flicker," he said in a
famous interview.
And what about the fact that just after the triumph of
the Moon, all three - Armstrong, Aldrin, and Collins -
resigned from NASA? Armstrong went to live in the

country, in Lebanon, Ohio, and did not give more interviews. When NASA celebrated the end of the Apollo program, he did not want to take part, and he also avoided the celebrations for the 40th anniversary of the landing in 2009. Moreover, in 1994, before a meeting of young people in which President Clinton was also present, Armstrong stated, "We still leave you many things to do[...], great turning points await those who will be able to remove the layers that 'protect' the truth." What did he mean?

Aldrin immediately returned to the Air Force but fell prey to depression, alcoholism, and drug addiction until he finally took his leave. He was a Freemason and was a member of the Lodge "Montclair Lodge No. 144" in New Jersey, and later affiliated with the Lodge "Clear Lake Lodge No. 1417" of Seabrook, Texas. He achieved the highest rank, Venerable Grand Master, the 33rd (the same rank that Licio Gelli had in Italy in P2).

Collins, who resigned from NASA in 1970, had the most remarkable career. He was called to the State Department as Assistant Secretary of State for Public Affairs. A year later, however, he was already 'just' the director of the National Air and Space Museum, a comfortable and quiet parking spot for veterans. Certainly not for someone who had been to the Moon.

16. Once upon a time, there was Apollo

"The word serves to hide the thought, the thought to hide the truth. And the truth strikes down those who dare to look at her"
Ennio Flaiano, A Martian in Rome

Vladimir Jakovlevič Propp can be considered the father of all storytellers. Ironically, he passed away on August 22, 1970, just four months after the "unfortunate" Apollo 13 mission returned to Earth. Propp, a professor of German origin born in Leningrad (now St. Petersburg) and passionate about the Russian language, wrote the seminal work "Morphology of the fairy tale" in 1928. In this work, Propp asserts that all fairy tales have common elements, regardless of their place of origin or the culture that created them. He goes on to define and illustrate the 31 characteristics present in fairy tales. For our purposes, only 13 are of interest, as they are fundamental:

1. The hero's departure from the safety of the initial environment creates tension and sets the story in motion.
2. The hero is prohibited from taking certain actions.
3. An antagonist enters the story.
4. The antagonist causes damage or a character lacks something important.
5. The hero reacts and decides to act to resolve the lack, which sets the conditions for future action.
6. The hero must overcome a difficult task or ordeal.
7. The hero acquires a magic object.
8. The hero and the antagonist engage in a direct confrontation.

9. The hero is freed from damage or the lack is solved.
10. The hero successfully completes the test.
11. The hero defeats the antagonist.
12. The hero is saved.
13. The hero returns and the story ends happily.

It is indispensable, therefore, for a fable to have an edifying morality that makes the reader/listener/viewer feel better (in this case, mostly the viewer). After the remarkable success of the Apollo 11 landing, NASA observed what Ennio Flaiano had already predicted in his story "A Martian in Rome" in 1954: the Martian who landed in the Capital was soon forgotten and even mocked by everyone after an initial period of astonishment on the part of the citizens.

Apollo 12 did not receive the same amount of attention as the previous mission. It is challenging for anyone to recall the names of the three astronauts on board (Charles Conrad, Richard Gordon, and Alan Bean). The mission that took place just four months before did not generate any headlines, and the '12' gained some attention only because the rocket was struck by two lightning bolts that endangered the lives of the astronauts and a significant portion of NASA's budget. Upon returning, it was rumored that the three astronauts had seen UFOs three times during their 'lunar' mission in order to gain attention for the Apollo project.

To regain the center of the world media scene and justify the insane public financing of the Apollo project, perhaps a fairy tale could be invented. Let us examine the elements of the Apollo 13 mission based on Propp's proposed analysis of the elements of the fairy tale. The astronauts, designated with the unfortunate number 13, leave for a risky and unknown mission. The ban on performing a certain action is precisely starting a mission with the forbidden number 13.

The launch was initially scheduled for Monday, April 13th (but started on the 11th) at 2:13 am, and the incident happened precisely on the 13th. Moreover, the capsule was named 'Odyssey' to generate more anxiety and attention among Americans. The infringement or antagonist is not a person but a fatality - the breaking of the oxygen tank. The damage is the concrete possibility that the three astronauts could perish in orbit due to the accident.

The reaction consists of all the actions put in place by the three to remedy the dangerous situation, which eventually makes them "Heroes." The proof is the one requested of them with the help of land in downtown Houston to solve increasingly complex problems. The resolution or magical object is the Lem, which becomes the "safe place" to take refuge and avoid death in space.

The struggle with the antagonist in this case is constituted by all the practical tests, which are reconstructed in the film 'Apollo 13' by Ron Howard, and engage the astronauts to transform the Lem lunar module into a module of re-entry on Earth. The removal is the release of the faulty service module four hours before returning home and that of the same Lem, which according to the fairy tale, will be "dropped" an hour before returning to the atmosphere. The solution consists of the victory of the Heroes in this series of complex technical-scientific maneuvers. The victory obviously comes after all these series of maneuvers and consists of a happy landing in the Pacific Ocean. The rescue is the recovery of the heroes by the helicopter departing from the Iwo Jima aircraft carrier at 1:07 on 17 April. The happy ending is the safe return of the Heroes home. The Good (USA) triumphed this time too.

As we have seen, the narration of the Apollo 13 mission fully falls within the canons of a fable. Thanks to all the suspense generated by the story, NASA could finally breathe a sigh of relief: the Apollo mission was once again on the front pages of all the newspapers in the world, and there was no need to fear any cuts in the financing of the project by Washington.

But there are also those who propose an even more radical reading of the fairy tale, even if the final result does not change. This is who investigated the Apollo 13 mission through the analysis of archive news from the Russian press, discovering that on April 8, 1970, in the Bay of Biscay, in the Atlantic Ocean, between France and Spain, a Soviet nuclear submarine caught fire. Due to the prohibitive weather conditions, searches continued until July 12th. Meanwhile, as we have seen, on April 11, the Americans had started the Apollo 13 mission, and the Russians - who were looking for their own submarine in stormy waters - also found, to their great surprise, a fake Apollo capsule without thermal protection, which was returned to the Americans only on September 8th, 1970.

What was the fake Apollo capsule doing at sea, so far from the NASA area of expertise? The story, which had remained unknown to the Western public for 40 years, was recently recounted and documented by Mark Wade, the director and founder of the Astronautics Encyclopedia. The story came to light when Nandor Schuminszky, a Hungarian space travel history enthusiast, sent Wade an incredible photograph found in a Hungarian newspaper from the 1970s. The caption read: "Murmansk: an Apollo capsule is handed over to some American delegates. [The capsule] was recovered by Soviet fishermen in the Bay of Biscay."

The capsule was then loaded onto the "Southwind," a US Coast Guard ship, to be taken home. Intrigued by the story, Wade contacted Schuminszky to learn more, as the story was not present in NASA's registers nor known to Western media. According to reports from the Russian site Novosti-kosmonavtiki, the Soviet experts who examined the capsule declared: "It was a model in thick galvanized steel, reconstructed very accurately and free of signs of corrosion. The weight, dimensions, and configuration of the control module were those of the Apollo capsules. [With the exception of] a searchlight beacon [...] and the fact that heat shields were not present. Everything was very simplified." The Americans called these exercise models "boilerplates" and used them frequently. For example, the BP-1204 capsule was used for exercises in Rota (in Spain), BP-1205 in Yokosuka (in Japan), BP-1223 in the Azores Islands, and so on. However, there is nothing about the original mission of the Boilerplate BP-1227, the capsule recovered by the Soviets in the Bay of Biscay and then returned to the Americans.

This story raises several considerations. First, it is evident that there was a silent agreement between Soviet and American authorities on the Apollo program. Second, the affair was embarrassing for the US, as neither American nor other Western media mentioned it. Third, it is possible that the embarrassment was related to the failed mission of Apollo 13, the only Apollo mission of 1970.

The relations between Apollo 13 and BP-1227 remain unclear. The capsule was returned to the Americans on September 8, 1970. However, it is uncertain when the Soviets recovered the capsule, by whom precisely, how, and under what circumstances. It is known that the capsule was recovered by a Soviet "fishing boat" (APATIT), but it is doubtful that the vessel was just a fishing boat, as military naval maneuvers were taking place in the area. Most likely, it was a spy ship. Wade does not mention the date of the recovery, which is unusual for navy records. This suggests that indicating the date of recovery could represent a source of embarrassment or national security risk for the United States.

Assuming that these events occurred on the night of April 11-12, 1970, just a few hours after the launch of the Apollo 13 mission from Cape Canaveral at 7:13 pm on April 11, it is possible that the capsule intercepted by the Soviets was not a mere training object, but the actual Apollo 13 capsule, which had been launched a few hours earlier for the fake lunar mission that would captivate the world for several days.

17. Vulgus vult decipi, ergo decipiatur
"The people want to be deceived, so let them be deceived", Cardinal Carlo Carafa (1517 - 1561)

To recap, on April 8, 1970, the Soviet nuclear submarine K-8 caught fire in the Bay of Biscay during the "Okean-70" naval exercises and was due to return to base on April 10. Rescue attempts were made until April 12, when the K-8 sank over 4,800 meters deep, resulting in the deaths of 52 Soviet sailors and 73 survivors. According to historical sources, the rescue attempts took place in stormy conditions.

The presence of many Soviet ships in the Bay of Biscay due to the K-8 tragedy could explain the interception of the BP-1227. Furthermore, if the weather conditions on the night of April 11-12 were as severe as reported, it would explain why American planes and helicopters stationed in Great Britain did not prevent the Soviets from taking possession of the capsule.

Continuing with this supposition, the capture of the capsule by the Soviets could have taken place only a few hours after the launch of Apollo 13, which could explain the reluctance of the American media to discuss the incident. American reconnaissance aircraft only appeared in the Bay of Biscay on the morning of April 12, searching for the capsule, which had since been recovered by Soviet ships. Eventually, an agreement was reached that returned the capsule to the US in September 1970. If this incident had occurred during any other period, such as the one reported on Wikipedia under "Boilerplate" (the recovery of 1227 in June 1969), the Americans could have forced any unarmed Soviet spy ships to immediately return the Apollo capsule. Additionally, there would have been no Soviet naval exercises taking place. Just 10 months prior, during the launch of Apollo 11, the American fleet had successfully repelled some disguised Soviet spy ships (posing as fishing boats) off the coast of Florida. However, during the military maneuvers "Okean-70," the unfortunate capture of BP-1227 occurred too far from home. It is likely that the events took place on the night of April 11th or 12th, 1970. The world was on edge as three astronauts were in danger of death, and their spacecraft, believed to be traveling to the moon, had been diverted to the Soviet port of Murmansk. According to NASA, Apollo 13 completed an orbit and a half around Earth before the third stage engine was reactivated to put it on a transfer orbit to the moon, which would allow for the necessary "slingshot" thrust to return to Earth.

However, Alexander Ivanovich Popov, author of "Americans on the Moon: A Great Breakthrough or Space Scam?" disagrees. He suggests that the Apollo spacecraft never even entered orbit, and that the rocket's trajectory was altered to fly within 100 kilometers of altitude and 300 to 700 kilometers from the American coasts. Popov claims that the Americans did not have the capability to develop a rocket powerful enough to carry the necessary equipment for lunar missions, and instead modified the Saturn-1 model into a more modern Saturn-1B. The Saturn-1B had a range of no more than 15 tons, and was covered with a heavy coating to make it appear more powerful. This coating, however, was so heavy that the rocket could not enter orbit.

The capsules transported, therefore, would have been free of astronauts and weighed no more than a ton, with a wall thickness of about 5 millimeters. The purpose of the rocket was to discreetly transport the equipment far away and out of sight, so that the "recovery of the capsule" could be staged at a later time.

Whatever the explanation for the story of Apollo 13, it is reasonable to doubt that NASA has told the whole truth for more than half a century.